THE ELECTRIC VEHICLE CONVERSION HANDBOOK

How to Convert Cars, Trucks, Motorcycles, and Bicycles

Includes EV Components, Kits, and Project Vehicles

Mark Warner, PE

HPBooks

HPBooks
Published by the Penguin Group
Penguin Group (USA) Inc.
375 Hudson Street, New York, New York 10014, USA
Penguin Group (Canada), 90 Eglinton Avenue East, Suite 700, Toronto, Ontario M4P 2Y3, Canada
(a division of Pearson Penguin Canada Inc.)
Penguin Books Ltd., 80 Strand, London WC2R 0RL, England
Penguin Group Ireland, 25 St. Stephen's Green, Dublin 2, Ireland (a division of Penguin Books Ltd.)
Penguin Group (Australia), 250 Camberwell Road, Camberwell, Victoria 3124, Australia
(a division of Pearson Australia Group Pty. Ltd.)
Penguin Books India Pvt. Ltd., 11 Community Centre, Panchsheel Park, New Delhi—110 017, India
Penguin Group (NZ), 67 Apollo Drive, Rosedale, Auckland 0632, New Zealand
(a division of Pearson New Zealand Ltd.)
Penguin Books (South Africa) (Pty.) Ltd., 24 Sturdee Avenue, Rosebank, Johannesburg 2196, South Africa

Penguin Books Ltd., Registered Offices: 80 Strand, London WC2R 0RL, England

THE ELECTRIC VEHICLE CONVERSION HANDBOOK

Front cover illustration by Pete Marenfeld
Back cover and interior photos by author unless otherwise noted

First edition: June 2011

ISBN: 978-1-55788-568-5

PRINTED IN THE UNITED STATES OF AMERICA

10 9 8 7 6 5 4 3 2 1

CONTENTS

ACKNOWLEDGMENTS

It would not have been possible to write this book without the help of a large number of very talented professional electric vehicle experts, enthusiasts, and backyard builders. First and foremost, I'd like to express my thanks to all the contributors of technical information, photos, illustrations, and drawings throughout the book. In no particular order of importance, my deep appreciation goes out to: Bob Batson of Electric Vehicles of America; Brian Berrett of Wilderness EV; Ryan Bohm of EVSource.com; David Brandt; Isidor Buchmann of BatteryUniversity.com; Mark Bush; Anthony Carstensen of Masterflux/Tecumseh Products Company; Mathew Clark, Carl Clark, and Spencer Stromberg of the Electric Car Company of Utah; Mike Dieroff of D&D Motor Systems; Rush Dougherty of EAAEV.org; Pearson Scott Foresman; Mark Gelbien of Enertrac Corporation; Chip Gribben of EVA/DC; Hannah Hamstra of NetGain Motors; Ed Matula of eRoadsters.com; Joe Miller of *Home Power* magazine; Dennis Palatov of DPCars.net; Li Ping of Ping Batteries; Dwayne Porter of U.S. Battery; Tony Helmholdt and his excellent ExperimentalEV.wordpress.com blog; Randy Holmquist and Bonnie Collins of Canadian Electric Vehicles; Kalyan Jana of Enersys; Peter Lee of Everspring; Bill Provence of PureElectricVehicles.com; Danny and Lysa Ray of Amped Bikes; Wistar Rhoads of KTA Services; Kristin Rogers of Dig Communications (Optima Batteries); Ryan Sberna of Eagle Eye Power Solutions; Bob Simpson; Roland Wiench; and Connie Yaskowski of East Penn Manufacturing/Deka Batteries.

For helping this mechanical engineer understand the magic of electricity, I'd like to thank: L.W. Brittian, Lonnie Cole, Bob Froehlich, Larry Krengel, Dr. Ken Mellendorf, Dr. John R. O'Malley of the University of Florida, Dr. Ralph J. Smith of Stanford University, and Burr Zimmerman.

For letting me lurk (and learn) from their EVDL posts, my deep appreciation goes out to: David Roden (EVDL Administrator), Bob Bath, Lee Hart, Jeff Major, Jack Murray, George Tyler, and Roland Wiench. Similarly, I've learned more than I ever hoped by the diy_ev_cars group at Yahoo groups postings of: Ronald Anderson, Craig Carmichael, Dave Cover, Ron Gompertz, Peter Eggleston, Don Eyermann, Nikki Gordon-Bloomfield, John Kester, Andy Laurence, Stuart Perkins, Gary Krysztopik, Ed Thaxton, Russ Sciville, Tom Simenatad, Pete Snidal, and Dan Wolstenholme. Believe it or not, your debates, arguments, and ever-so-tangential ramblings made me think about and understand electrons and EVs at a much higher level.

I also need to express my heartfelt thanks to the owners of the project vehicles featured in this book. You inveterate tinkerers, DIYers, and clever inventors are the perfect examples of the spirit of the EV community. Many, many thanks go out to: Tim Catellier, Mark Hayes, Wayne Krauth, Rob Nicol, David Oberlander, James Parsons, Edward Reims (and Mike Sileo for the photos), Lennon Rodgers, and Fred Weber. I am still amazed at the quality and performance of your individual EVs. You guys are an inspiration to the rest of us.

Jack Rickard and his excellent EVTV videos deserve a special shout-out here, too. Thanks for creating and sharing your extensive knowledge with the rest of us.

My deep gratitude also goes to Alan Adams, Lee Hart, Bob Bath, Gary Heinz, and Roger Werner for providing invaluable review, technical feedback, and input to the manuscript. This book is better because of your help and expertise.

Of course I would be remiss if I didn't mention my publisher, Michael Lutfy of HPBooks. Thank you once again for the opportunity to research and write about a subject I enjoy.

Finally, I want to thank my beautiful wife, Sonya for her support and patience while I slogged my way through another manuscript. *Je t'aime.*

INTRODUCTION

"Failing to plan is planning to fail."—A. Lakein

This is a book that describes how to convert an internal combustion-powered car or truck into an electric vehicle (EV). At first glance, this may seem like a daunting task for someone new to the world of EVs, but rest assured that converting a car to electric power isn't nearly as difficult as it may seem. In fact, the process is rather straightforward if you have taken the time to learn all the ins and outs and dos and don'ts beforehand. When it comes to tackling an EV conversion project, knowledge is power, which is why this book was written by an experienced professional engineer but targeted toward the amateur enthusiast and written in the language of the layman. The concepts and procedures outlined herein are not difficult to understand, nor are they too challenging for the average person to undertake. EV conversions have been successfully completed by 14-year-old amateurs, by 88-year-old retirees, and just about everyone else in between. If these people can do it, then so can you—and this book will help you understand how.

When planning this book, we wanted you, the reader, to understand why one should (or should not, as the case may be) convert a car to electric power. We wanted you to grasp the fundamentals of electricity and then be able to apply them to powering a car. And we wanted you to understand every step, nuance, and phase of the conversion process ahead of time—before you pick up the first wrench or shop for the first wire or battery. Converting your car or truck to electric power isn't very hard if you learn the basics ahead of time.

With that in mind, this book is organized into sections to help you understand the process in a step-by-step manner.

The first part of the book begins with an explanation of what an EV is (and what it isn't) and how EVs are constructed and operate. We then explain the pros and cons of electric vehicle ownership, including a look at some of the more common myths, rumors, and untruths surrounding electric vehicles. We then lay out a series of questions that will help you decide whether an EV is the right type of vehicle for you. We also describe the step-by-step process of converting an internal combustion engine (ICE) vehicle to electric power. Finally, we discuss the merits of building your own EV versus simply buying a preexisting electric car. EV construction and/or ownership is not for everyone and understanding this fact before committing to a conversion project is an important first step.

The next part contains the majority of the technical information on EV conversions. It's where we go through in detail all of the components and parts required to convert a car or truck to electric power. This section is subdivided into seven chapters, covering motors, controllers, batteries, chargers, wiring, accessories, and instrumentation. In each of these chapters we not only cover component selection, but also the theory of operation and practical considerations to keep in mind—again, to help you make the most informed choices at the beginning of the process, rather than when it's either too late or too expensive to change your mind.

After learning the ins and outs of the conversion process, the Project Vehicles section will hopefully be a source of inspiration to you. In this section we show how nine ordinary enthusiasts converted their own vehicles to electric power. The vehicles we look at here range from simple bicycles and motorcycles to moderately complex commuter cars to sophisticated sports cars that will completely change your notion of what an electric vehicle is capable of doing on the street.

In the appendices, we cover a number of more advanced topics, including the concepts of torque, work, energy, and power. This section also delves into the basics of electricity and electrical circuits. The answers to questions like "What are voltage and current?" and "How does electricity flow in a wire?" are provided here. Understanding these concepts and principles isn't strictly necessary to converting a car to electric power, but it will significantly help you make more informed decisions and choices when in the midst of a conversion process.

Here we also touch on the issue of EV safety, including how to design, build, operate, and maintain an electric vehicle in a safe and responsible manner. **Please ensure that you read this section before, during, and after reading the rest of the book.**

Included in the final sections of the book is a comprehensive glossary of technical terms and phrases used in the EV world. Many of the problems beginners encounter when learning about EVs can be traced to simply not understanding the language of the subject. This glossary should give you a confident leg up on the learning curve of electric vehicles.

Wrapping up the book is a list of popular EV suppliers and reputable sources of information to consult when planning your own conversion project.

Finally, when you've finished reading this book, we encourage you to examine other sources of information on the subject of EVs. Seek out information from EV clubs and other enthusiasts in your area. Read EV forums, search the Web, talk to experts. Knowledge is power, and never has this been truer than when it comes to converting a car to electric power. There is a wealth of wisdom, data, and knowledge available to you just for the asking. Don't be shy; building an electric vehicle may seem like rocket science at first, but it's really not. In fact, it's relatively easy—if you take the time to learn the basics. And then, not only will it be easy but it will also be fun and rewarding, save you money, and just might help rescue the planet at the same time. Aren't these the reasons that you picked up this book in the first place?

Good luck and happy (electric) motoring!

—Mark Warner, PE

DEDICATION

This book is dedicated to my two perfectly imperfect children, Niko and Katya.

GETTING STARTED

Chapter 1

What Is an Electric Vehicle?

"Knowledge is a process of piling up facts; wisdom lies in their simplification."
— Martin Fischer

While they may look and operate very differently from each other, the four vehicles shown here are all electric vehicles, or EVs; they each have an electric motor that propels the vehicle. Shown clockwise from the upper left corner are: 1) an electric trolley car that receives electric power from overhead lines; 2) a diesel-electric locomotive engine that creates its own electricity from an onboard diesel-powered generator; 3) a hybrid car that carries onboard batteries that are recharged from an onboard internal combustion engine (ICE) and generator; and 4) a fuel-cell vehicle that creates power for its electric motor via the conversion of chemical energy into electricity through a device called a fuel cell.

Imagine never having to stop at a service station and pump gasoline into your car's tank again. Imagine driving to and from work for pennies per mile. Imagine driving to the grocery store in a silent, near-zero emissions vehicle that needs just a fraction of the yearly maintenance that a traditional car requires. Now, imagine a car that not only saves you money to operate, but helps the country lessen its dependence on foreign oil while providing a tax incentive to you at the same time. Congratulations: you've imagined the electric vehicle, or EV.

An EV is defined as any vehicle that gets some or all of its power from electrical energy. All EVs use some type of electric motor to provide propulsion, but there are many different ways that EVs can be supplied the electricity necessary to operate the motor. Some EVs, such as trolley cars, use electricity that is provided directly from a power station by way of transmission lines; electricity is delivered to the vehicle via brushes that contact overhead power lines. Others, such as diesel electric trains, create their own electricity by way of an onboard generator that is driven by a large internal combustion engine, or ICE. Other EVs, like hybrids, have both a traditional internal combustion engine onboard and an electric motor with batteries. Hybrids switch automatically between ICE and electric power (and sometimes even work with both systems operating together) depending on driving conditions and load. Still other EVs, such as fuel-cell vehicles, are able to convert gasoline or hydrogen directly into electricity via a chemical process. While certainly interesting, none of these types of EVs are the subject of this book. Instead, we will focus on so-called battery electric vehicles, or BEV.

A BEV is an electric vehicle fitted with batteries that it carries onboard to supply the electricity needed for the motor to work. Like the gas tank of a conventional ICE-powered vehicle, batteries in a BEV are simply an energy storage device. An external electrical supply source, such as an electrical outlet in a garage, is periodically connected to the vehicle and used to charge up the batteries. Once the batteries are charged, the vehicle can be disconnected from the external power source and driven normally around town or on the highway. When the driver wants to go somewhere, he or she uses the energy stored in the batteries to power the electric motor and, consequently, drives on the road like any other car or truck.

In the first part of this book, we will discuss the pros and cons—and common myths—of BEV ownership, including the different options of acquiring a vehicle, such as buying versus converting an existing ICE vehicle to electric power. This latter item, by the way, is probably why you purchased this book in the first place, and it will be our focus in most of the subsequent chapters. But before we do any of these things, let's stop and examine the major parts of a typical

One of the most common types of BEVs is the ubiquitous golf cart found on golf courses and the streets of retirement villages. Fundamentally, an electric golf cart is no different than a more advanced BEV that is constructed by converting a gasoline powered car or truck and used on public streets and highways. The primary differences between the two are in the size of the overall system voltage, the horsepower rating of the motor, and the capability of the motor controller.

Battery electric vehicles come in many varieties and types. The focus of this book is on road-going cars, trucks, motorcycles, and bicycles that can be converted to electric power. The BMW Z3 convertible shown here is an example of such an animal. It's an extremely well executed high-performance BEV that features an 11" DC motor, sophisticated motor speed controller, and advanced lithium-ion batteries. The roadster can be recharged in as little as four hours, can travel 60+ miles on a single charge, and can speed down the highway at over 80 mph. Courtesy Tim Catellier.

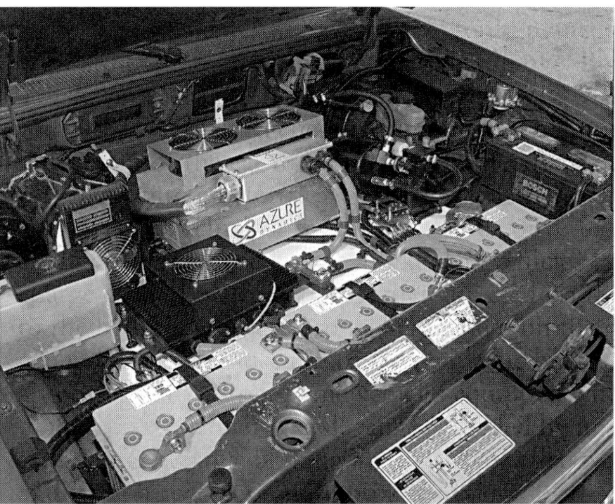

For the uninitiated, the under the hood components and inner workings of a BEV can be difficult to grasp at first. Rest assured, however, that there is nothing mysterious or particularly difficult to understand. By the time you've finished reading this book you should be able to plan, specify, and build your own electric vehicle. Courtesy Ed Reim.

battery electric vehicle and how they work together. **Note:** Throughout the remainder of this book we will use the terms *EV* and *BEV* interchangeably to mean a battery electric vehicle.)

How a Battery Electric Vehicle Works

Before starting, we need to ensure that you know a little bit about how electricity works. If you don't, Appendix A at the back of this book is a short and simple primer on basic electricity and how electrical circuits operate. If you're unsure at all about electricity or wiring diagrams, take the time now to review this material, as well as Appendix D on EV safety. Then come back here and pick up where you've left off.

To the uninitiated, battery electric vehicles appear complicated. Wires, controllers, electronics, cables, batteries—if you don't know what you're looking at, the collection of electrical components and circuits underneath the hood of a typical BEV can be both perplexing and intimidating. What are all those electro-mechanical things and how do they work?

To understand the components and operation of a typical BEV, let's start with an overly simplified BEV, and then add in the different components and subsystems of a modern, complete vehicle in a step-by-step manner.

The simplest battery electric vehicle we could build would be one in which an electric motor is connected directly to the drive wheels of the vehicle. In other words, imagine a car, truck, or motorcycle that has a large motor bolted directly to the rear wheels. Then imagine a large battery wired to the motor with a big disconnect switch installed in the line. When the operator of the vehicle wants to go forward, he or she would close the switch, which would complete the electric circuit and allow electricity to flow from the battery to the motor. The motor would turn, and, consequently, so would the drive wheels. Voilá! We've invented the electric vehicle.

Okay, so things aren't really that simple in the real world. For one thing, an electric motor that could be connected directly to the drive wheels of a

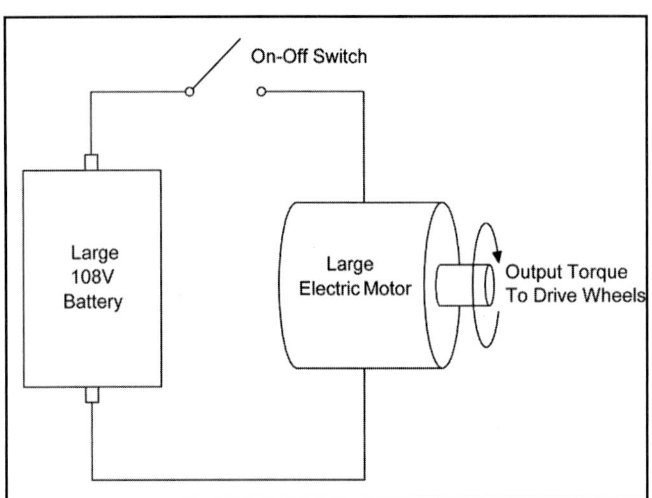

Building an EV—Step 1: To best understand how a BEV system is constructed and how all the parts and components work together, it's useful to start with a simplified BEV electrical schematic and build up from there. Shown here are the simplest of BEV circuits. A large 108V battery is connected to a large electric motor. Control of the motor is via a simple on-off switch. Some very rudimentary BEVs are actually constructed in this manner. For example, many home-built electric bicycles operate using a single large battery, a simple hub motor built into the wheel, and a trigger on-off switch.

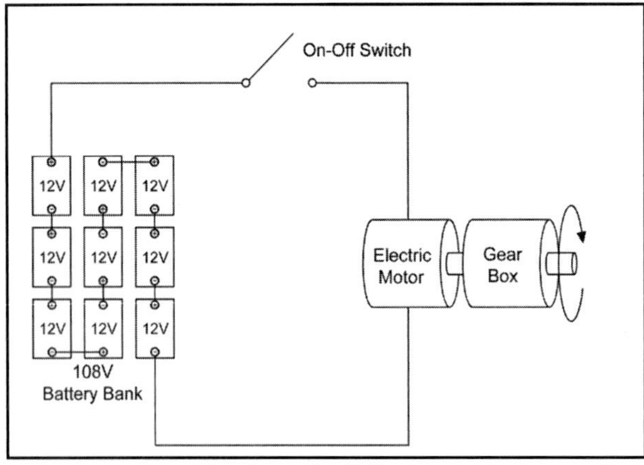

Step 2: While the simple circuit shown above has a place in some basic applications (e.g., a simple electric bicycle), it would not be a very practical solution for a car or truck. For example, it would be challenging to find a supplier that would sell you a single large 108V battery. Packaging it in the space constraints of a car or truck might also be difficult. The solution is to wire together a number of smaller 6- or 12-volt batteries to achieve the same overall system voltage. Similarly, a smaller electric motor, coupled to a gearbox or transmission, would be both less expensive and easier to mount in a car. Note that this circuit and the previous circuit are functionally identical to each other, but this one is less expensive and easier to package into a car or truck.

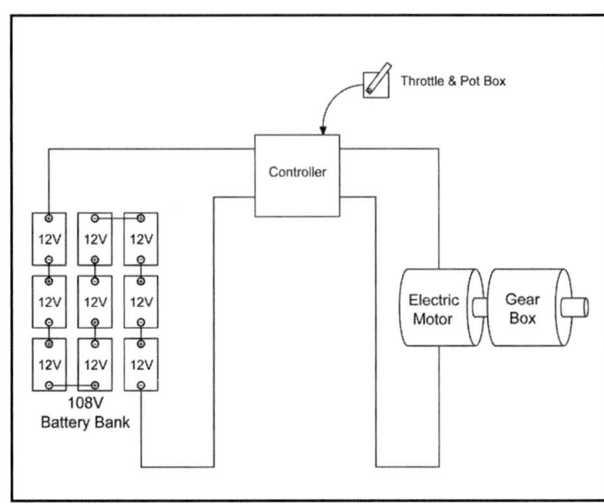

Step 3: The next step is to replace the on-off disconnect switch with something that provides a better means of smoothly and precisely controlling the speed of the motor. To accomplish this, we replace the switch with an electronic device called a "controller." We will cover the operation of controllers in a later chapter, but for now it's important to realize that the flow of electricity from the battery bank to the electric motor can be accurately varied by the operation of the controller. Note that a throttle and "pot box" is also connected to the controller; this is the input device that the operator of the BEV uses to tell the controller what speed he or she wants.

typical commuter car would have to be extremely large and powerful. In other words, it would be too big and expensive to be practical. So instead of a really big motor, we can substitute a smaller, less powerful (and less expensive) motor, along with a gear reduction box, or transmission. The transmission allows the smaller motor to do the work of a larger motor by trading motor speed for drive torque.

The next thing we would want to replace in our system is the one large battery with a series of smaller, less-expensive batteries that are wired together to act as one large unit. Again, the reason is primarily economics; it's cheaper to procure, say, ten relatively small 12-volt batteries and connect them together than it is to buy one large 120-volt battery. Packaging in small vehicles is also easier with multiple small batteries than one large unit.

The Controller—Next we turn our attention to the on-off disconnect switch. So far, our simple BEV has only two speeds: zero (off) and maximum (on). This clearly isn't desirable, as we need a means of controlling the speed of the motor and, hence, the vehicle itself. So, instead of a big on-off switch, we substitute an electronic device called a *controller*. There are a number of different types of controllers that operate in various ways, but they all do the same basic thing: They control the flow of electricity to the motor.

The Pot Box—Most modern controllers have something connected to them called a *pot box*, which is usually just a variable resistor that can be attached to the accelerator pedal. When the pedal is pushed down by the driver's foot, the resistance of the pot box changes accordingly. The electronics in the controller interpret this resistance change as a command to increase or decrease speed. The controller then correspondingly varies the amount

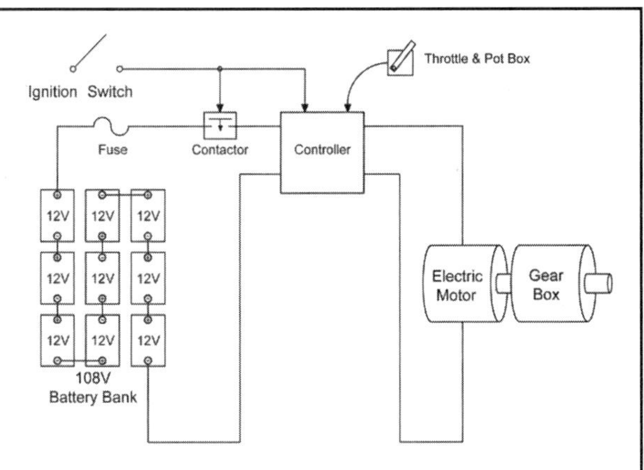

Step 4: Next we need a means of arming, or turning on and off the electrical system and the controller. For this, we will reinstall a large on-off switch we took out in the previous step, but we won't use a manually operated switch. Instead, we will use something called a solenoid contactor, which is an electronically operated on-off switch that is activated by a smaller switch or "ignition" key. A solenoid contactor keeps the operator from putting his or her hands directly on a large high-current switch to turn the system on or off. Instead, the operator is isolated from the high voltage, high-current main circuit via the low voltage, low current "ignition" circuit. Note that the ignition switch is also wired into the controller as an "enable" device.

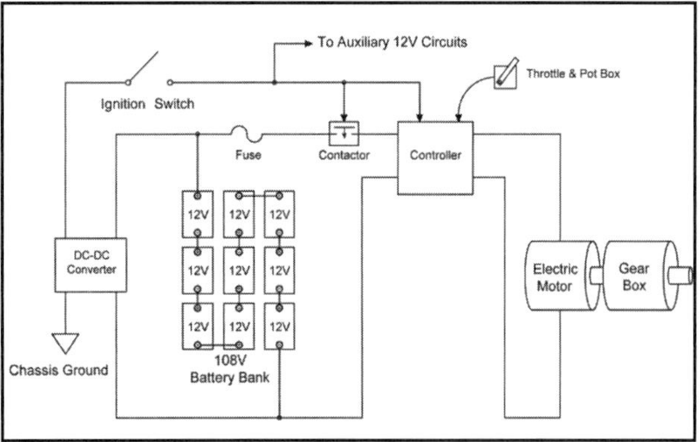

Step 5: To power all of the smaller 12-volt items on the car or truck (e.g., the lights, horn, radio, wipers, etc.) we add a device called a DC-to-DC converter. This electronic box is connected in parallel with the main battery bank, and simply steps down, or reduces the 108 volts to 12 volts.

of electricity that flows to the motor from the battery. If the accelerator pedal is pushed down further, the resistance changes in the pot box, and the controller responds by allowing even more electricity to reach the motor. When the accelerator pedal is released, the controller responds by stopping the flow of electricity.

On-Off Switch—Next we need a means of arming, or turning on and off the entire electrical system. For this, we will reinstall the big on-off switch we took out in the previous step. This time, however, we won't use a manually operated switch. Instead, we will use something called a *solenoid contactor*, which is nothing more than an electronically operated on-off switch that is activated by a smaller switch, or *ignition* key. This type of system is used on most BEVs for safety purposes; we don't want the operator putting his hands directly on a large high-current switch to turn the system on or off. Instead, we want to isolate the operator from the high-voltage, high-current main circuit, and we do so with a secondary ignition circuit.

We're getting close to a complete BEV system, but there are a few more things we have to add to the circuit to complete it. For instance, a fuse or circuit breaker in the main power system would be useful in the event of a short circuit or fault with the motor, controller, or batteries. Just like breakers or fuses used in your house, the circuit breaker in a BEV circuit is designed to trip or break if the current passing through it becomes too high. Fuses can provide the same type of protection, but they tend to be non-reusable; i.e., if a fuse "blows" due to high current, you have to replace it with a new one. Either fuses or circuit breakers are installed in most BEV systems.

We also need a means of measuring the amount of electricity in the battery and how much is flowing through the system at any given time. Instruments like this are not strictly needed to operate a BEV, but just how fuel level, oil pressure, and water temperature gauges are used in ICE vehicles to help the driver operate the vehicle and to alert him to fill up or perform routine maintenance, electrical gauges can be very useful in a BEV. The two most common instrumentation devices used on BEVs are ammeters, which measure current flow, and voltmeters, which measure voltage. Ammeters are usually installed either in series in the main circuit, or are used in conjunction with something called a *shunt*, which is wired in series with the main circuit. In contrast, voltmeters are typically installed in parallel with major circuits and are used to measure voltage potential differences between two points in the circuit, such as from one side of the battery bank to the other.

Converter—Next in our conversion project is the need to power other vehicle subsystems. The main electric drive motor needs high voltage to operate (typically 72 to 200 volts, or more), but all of the traditional vehicle accessories, such as headlights, horn, wipers, radio, and the like, typically operate on just 12 volts. To supply this lower voltage to these items, something called a DC-to-DC converter is often wired into the system. The converter "steps down" the high main system voltage to a nominal 12 volts for auxiliary use. On some BEVs, a small 12-volt auxiliary battery is sometimes used in place of the DC-to-DC converter. On other BEVs, a 12-volt battery is used

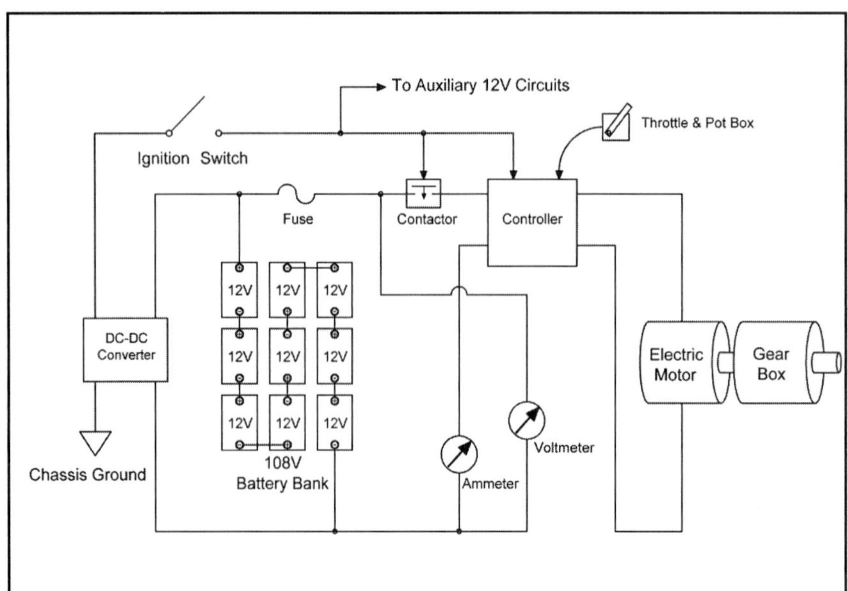

Step 6: We also need some means of measuring the overall voltage of the battery bank, as well as the amount of electrical current that is flowing to the motor. A voltmeter is wired in parallel with the battery bank, and therefore measures its overall system voltage. An ammeter is wired in series with one of the main feed wires to the controller, and therefore measures the instantaneous amperage flowing to/from the battery bank.

in conjunction with a DC-to-DC converter as a type of buffer, or reservoir, between the converter and the auxiliary vehicle systems.

Charger—Finally, to complete our basic BEV circuit layout, we need a means of recharging the batteries after they are depleted from use. The device used for this purpose is called a charger, and it converts household AC electrical power into DC power. Some BEVs carry a charger onboard the vehicle, while others leave the charger at home. The advantage of the former is that it allows "opportunity charging" of the batteries anywhere there is an electrical outlet that has sufficient voltage and amperage capabilities to recharge the batteries. For instance if a BEV is used for commuting to work, the batteries can be charged at home overnight for the trip to the office, and then recharged again during the day while the owner works. The batteries would then be ready for the return trip home. Some very lightweight vehicles and vehicles with high capacity batteries that don't need frequent recharging usually employ only a home-based charging system for use just overnight.

And that's all there really is to a battery electric vehicle—a battery bank, an electric motor, a controller, a fuse or circuit breaker, some gauges, a DC-to-DC converter, a charger, and some wiring. What could be simpler? Of course, the devil is in the details, as they say, and there are a number of trade-offs and complexities that need to be considered when planning a BEV conversion project. We will look at the subtleties and details of all the parts, components, and subsystems in later chapters. For now, though, you already know more than most people do about the operation of a battery electric vehicle. Next in your EV education is to look briefly at some of the more important advantages, disadvantages, and myths of EV ownership.

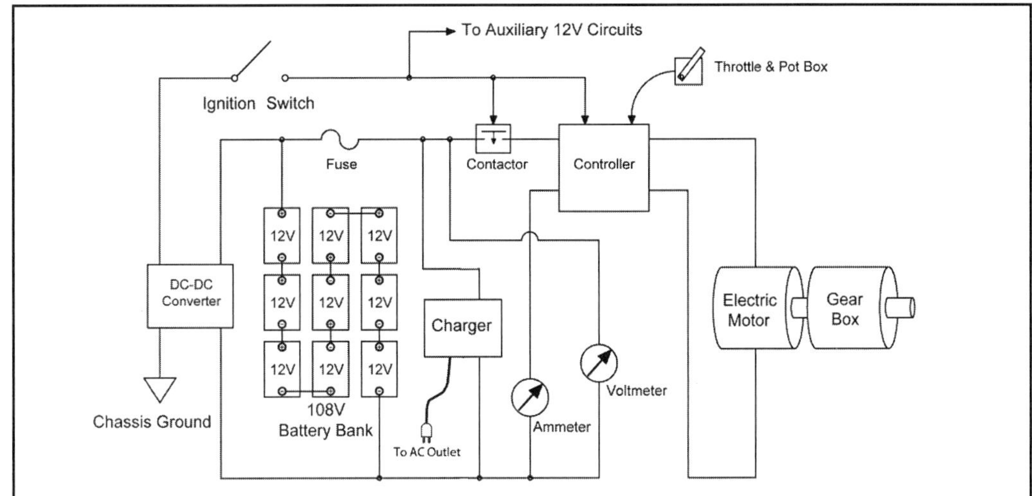

Step 7: The final piece of a circuit is the addition of a charger, which is used to convert household 110 or 240 volts AC into direct current that is fed back into the battery bank. This is the principle means by which the battery bank is recharged when the car is parked overnight in a garage.

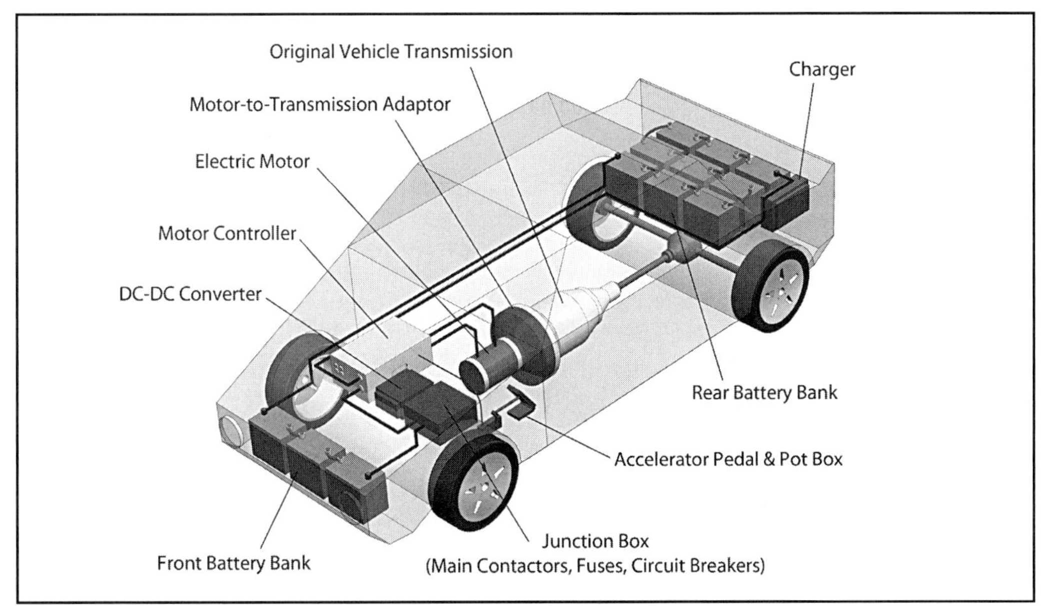

The major components of a BEV are shown in this image, including a front and rear battery bank, a motor and adapter connected to the original vehicle transmission, a speed controller, a throttle and pot box connected to the controller, a DC-to-DC converter, a large junction box that houses the main switch contactors, fuses, and circuit breakers, and an onboard charger mounted next to the rear battery bank. Note that only some of the major wiring is displayed, and that instrumentation (such as a voltmeter and/or ammeter) are not shown. In this particular vehicle, twelve 12-volt batteries are wired in series to provide a total system voltage of 144 volts. Underneath their body's sheet metal, nearly all battery electric vehicles are constructed in a similar manner as this vehicle.

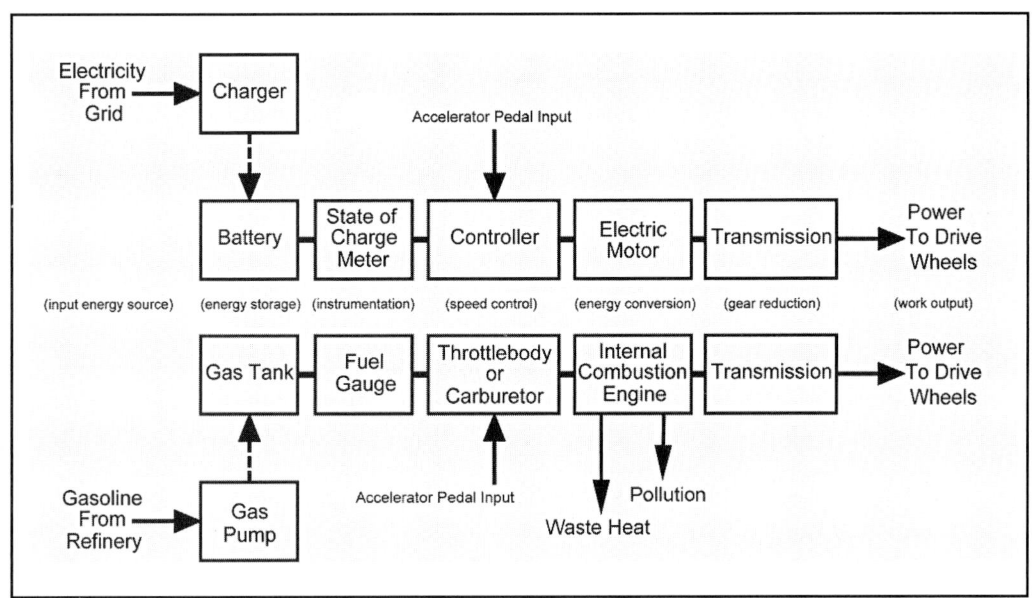

As shown here, there is a one-to-one comparison between the basic components and systems of an ICE-powered car or truck, and those of a battery electric vehicle. Energy is brought in on the left in the diagram and captured in some type of storage device. In the case of a BEV, the battery bank serves this purpose and is used to store input electrical energy from the local power grid. A motor controller is then used to control the amount of electricity delivered to the electric motor. The major difference between the BEV and the ICE is the amount of pollution and waste heat that the latter emits compared to the former.

Chapter 2

Is an EV Right for You?

"The only thing more expensive than education is ignorance."
—Benjamin Franklin

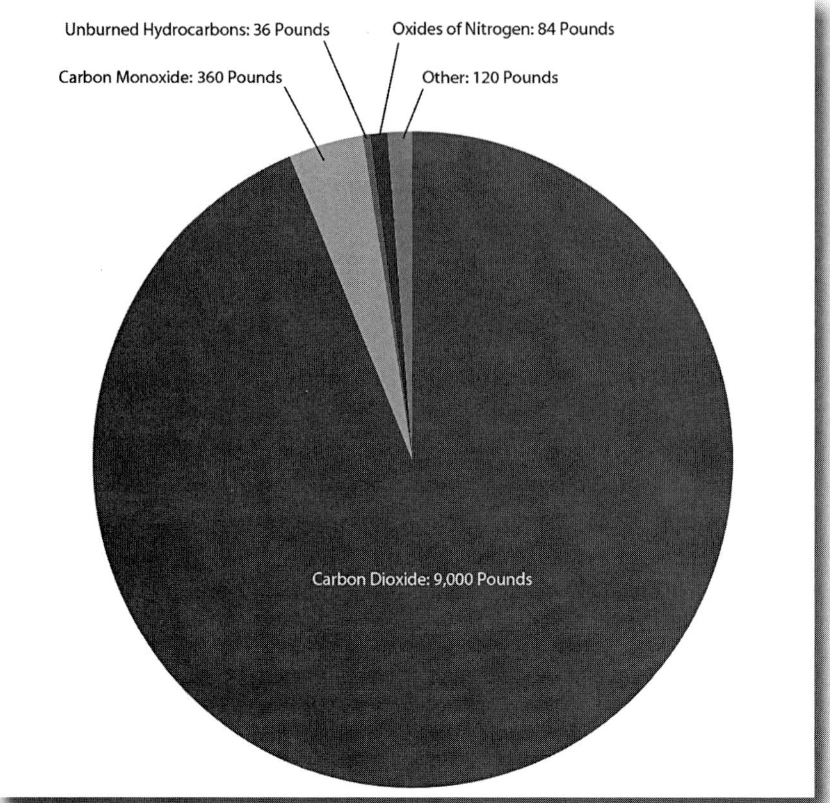

The average internal combustion powered vehicle emits nearly 4.5 tons of carbon dioxide into the air every year. Also spewed out the tailpipe are significant quantities of carbon monoxide, unburned hydrocarbons, nitrogen oxides, and a variety of other pollutants, including soot, particulates, sulphur oxides, and other nasty stuff.

Battery electric vehicles aren't right for everyone, and they may not be right for you, either. There are very compelling advantages to owning a BEV, but there are also some disadvantages. Worse, there are many myths and rumors about electric vehicles on the Internet, making the process of deciding whether an EV is appropriate for you a daunting and confusing task. Separating the pros from the cons, and the facts from the fictions, can be a difficult job for someone new to EVs. In this chapter we will take an uncompromising look at electric vehicle ownership, examining what's good about them, what's bad, and what is downright untrue. Let's start with the easy part first: the advantages of EV ownership.

Advantages

There are a variety of excellent reasons why you might consider owning an electric vehicle. These range from altruistic social responsibility reasons to purely egotistical motives of driving something cool and unique. For others, the most compelling reason to buy or build a BEV is rooted in basic economics; owning a battery electric vehicle can save the consumer significant money, both in operational expenses and maintenance costs. Still others choose EVs for less obvious benefits, such as reducing their time spent commuting or just being part of a unique and fascinating subculture of fellow EVers. Let's look at some of these more important advantages.

Social Responsibility and the Environment—There are a number of ways to state this, but perhaps the simplest is just this: BEVs are good for the environment. When driving on the road, the average battery electric vehicle produces essentially zero pollutants. Let's say that another way: if you drive just one mile in a typical ICE-powered vehicle, such as a late '90s Ford Focus or Honda Civic, approximately 0.03 lb of carbon monoxide, 0.003 lb of unburned hydrocarbons, 0.007 lb of nitrous oxides, and 0.75 lb of carbon dioxide are released from the tailpipe of the car into the atmosphere. These numbers may sound small, but the average American driver travels approximately 12,000 miles per year in their car. This means that just one commuter vehicle can emit four and a half tons of carbon dioxide into the air during the course of a single year.

Sounds awful, doesn't it? Well, it gets worse when you consider that there are more than two hundred million passenger vehicles (yes, 200,000,000!) on the road in the United States. This is 2.4 trillion commuter miles, or the equivalent of five million round trip visits from the earth to the moon every single year! And this ignores all the other vehicles in North America, South America, Europe, Asia, Australia, and Africa. Oh, and let's not forget that more than half of all vehicle miles driven are in cities and towns, where the majority of driving is in stop-and-go traffic.

Gasoline- and diesel-powered vehicles do not switch off

whenever they are stopped at a light. They continue to burn fuel, produce pollutants, and generate noise. In 2003, more than 2.3 billion gallons of fuel were wasted idling in traffic jams and stoplights in the U.S. In comparison, a battery electric vehicle, when at rest at a stoplight or in heavy traffic, is essentially completely off, consuming zero amps of electricity. When the light turns green or the traffic jam clears, the accelerator pedal is pressed down and the amps start flowing again. Whenever the car is stopped, however, zero energy is consumed by the motor and zero pollution created.

Now, astute readers will note that the electricity to charge a BEV's batteries has to come from somewhere. Often this source is a coal- or oil-fired power plant. While it's true that these behemoths produce more than their fair share of pollution, the amount of emissions released into the air that is associated with the electricity used to charge a set of batteries is far less than an equivalent ICE-powered vehicle driven the same number of miles. This is due to a variety of reasons. For example, unlike automotive gasoline-fueled engines that have to operate over a very wide—and therefore inefficient—speed range, electrical power plant generators are designed to run at near constant speeds that maximize their efficiency.

Chip Gribben's eye-opening study *Debunking the Myth of EVs and Smokestacks* found that EVs would be cleaner than ICE vehicles even if all of the electricity they used was produced by coal-fired power plants. Since half of the U.S.'s electric power is actually produced by cleaner sources than coal, EVs are essentially twice as good as this. More important, as older power plants are shut down, upgraded, or replaced by cleaner/renewable sources, the pollution output per kilowatt of electricity will drop even further.

Another factor to consider is that automotive engines are notoriously inefficient, throwing away nearly 75% of the energy released during combustion as waste heat, either out the tailpipe or into the cooling system. While not perfect by any means, modern electrical power plants are still roughly twice this efficient. They are also always on, which means there is no warm-up period (which in ICE-powered vehicles is often the time of greatest exhaust emissions). There are also economies of scale that factor in, including the emissions control systems. Large power plants are fitted with large, efficient emissions control systems. Automobile engines are usually equipped with the minimal systems allowed by law (for reasons of both cost and weight).

Power plants are also highly regulated by federal

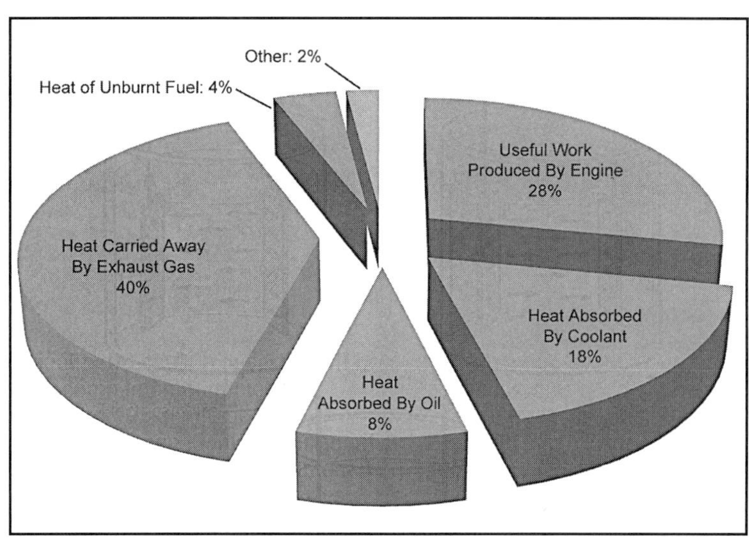

Internal combustion engines (ICE) are extremely inefficient. For every one BTU of energy converted to useful work, more than two BTUs are thrown away as some type of heat or pollution.

and state laws, and are routinely checked for levels of harmful emissions; many municipalities in the U.S. don't even require annual vehicle emissions testing, and those that do often only test once every two or three years. And let's not forget that ICE-powered cars and trucks produce the least pollutants when they're new; as they age and their internal rings and seals wear, the pollution output increases.

Even more compelling is the fact that that many newer electrical generating stations being constructed are essentially non-polluting during operations. These stations are designed to use so-called "renewable" forms of energy. Power plants that fall into this category include wind turbines, geothermal, photovoltaic solar, solar-thermal, hydroelectric, and even tidal-based systems. We also can include nuclear power plants in this list. "Nukes" don't use renewable energy sources, per se, and admittedly have long-term waste storage problems that have yet to be solved permanently, but they operate in a very clean manner and release no harmful emissions into the air during normal operations.

Electricity can be generated in many different and clean ways. In contrast, creating gasoline and diesel is possible only via refining of crude oil. To create a gallon of gasoline, a well has to be drilled, oil pumped out, refined and distilled, and the resulting fuels have to be shipped, trucked, and/or piped to fueling stations. All throughout this process, fumes and out-gassing is taking place, including the simple process of a consumer adding gas to his or her tank at a convenience store pumping station. Then, when the ICE-vehicle is operated, all kinds

Gasoline and diesel are refined from petroleum crude oil, which is a toxic and flammable liquid. Unfortunately, crude oil is found under the surface of the earth and must be drilled and pumped out. Disasters like the Exxon Valdez tanker spill in Alaska, and the BP Deepwater Horizon drilling rig spill in the Gulf of Mexico are proof that the drilling, pumping, and transport of the oil presents a serious risk to the earth's ecosystem. The act of refining the oil into gasoline is also fraught with pollution hazards, as is the piping and trucking of the product to your local filling station.

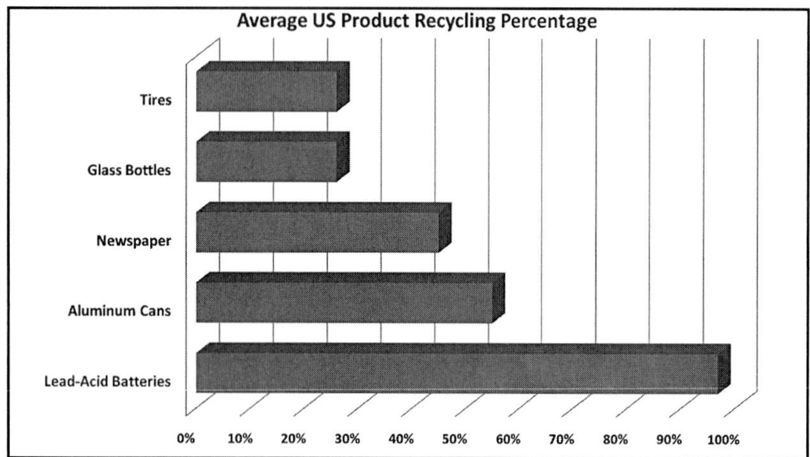

Lead-acid batteries are one of the most recyclable products in use today. More than 98% of the materials in a battery can be recycled and reused.

While it's true that over half of the electricity in the U.S. is created by burning coal, that percentage is dropping on a yearly basis. Unlike gasoline, which requires the refining of petroleum, electricity can be created in a wide variety of different ways. Clean, renewable sources such as hydroelectric, wind, and solar are becoming more and more prevalent.

of harmful byproducts are created and released into the air we all breathe. Even when compared to the next-generation of ultra low exhaust ICE-vehicles, battery electric cars, trucks, and motorcycles are much, much cleaner and environmentally friendly.

BEV vs. ICE Economics—While environmental considerations are the reason many people own and operate BEVs, an equally large percentage of owners jump on the EV bandwagon for a more selfish reason: economics. Saving the environment is important, but so is saving money. Electric vehicle purchase costs are roughly the same up front as a traditional ICE-powered vehicle, or even a little more, and converting an old ICE car to electric power can be done for approximately the same amount of money as rebuilding its original engine and restoring its subsystems. So why consider an electric vehicle if it's going to cost roughly the same as a comparable ICE vehicle? Answer: The long-term costs of ownership are lower for the BEV. The fuel costs are lower, maintenance is less, and repairs are fewer. If you keep a BEV for more than a few years, it will come out ahead of an ICE in costs. Let's look at this a little closer.

While a few owners of battery electric vehicles have their own standalone solar or wind-power systems (and can therefore operate essentially for free), most owners use electricity supplied by a local utility company to charge their vehicle's onboard battery banks. Typically the owner charges their vehicle overnight while he or she sleeps, and then drives to work, school, or shopping during the day. When the vehicle is returned home at the end of the day, the batteries are plugged back into the charging system for replenishment. By the time the car is needed the next morning, the batteries are charged up again.

So how much does it cost to recharge a battery bank overnight, and, more important, how much does it cost per mile to operate a BEV, and how does this compare to a traditional ICE-powered vehicle? To calculate these things, we need to first

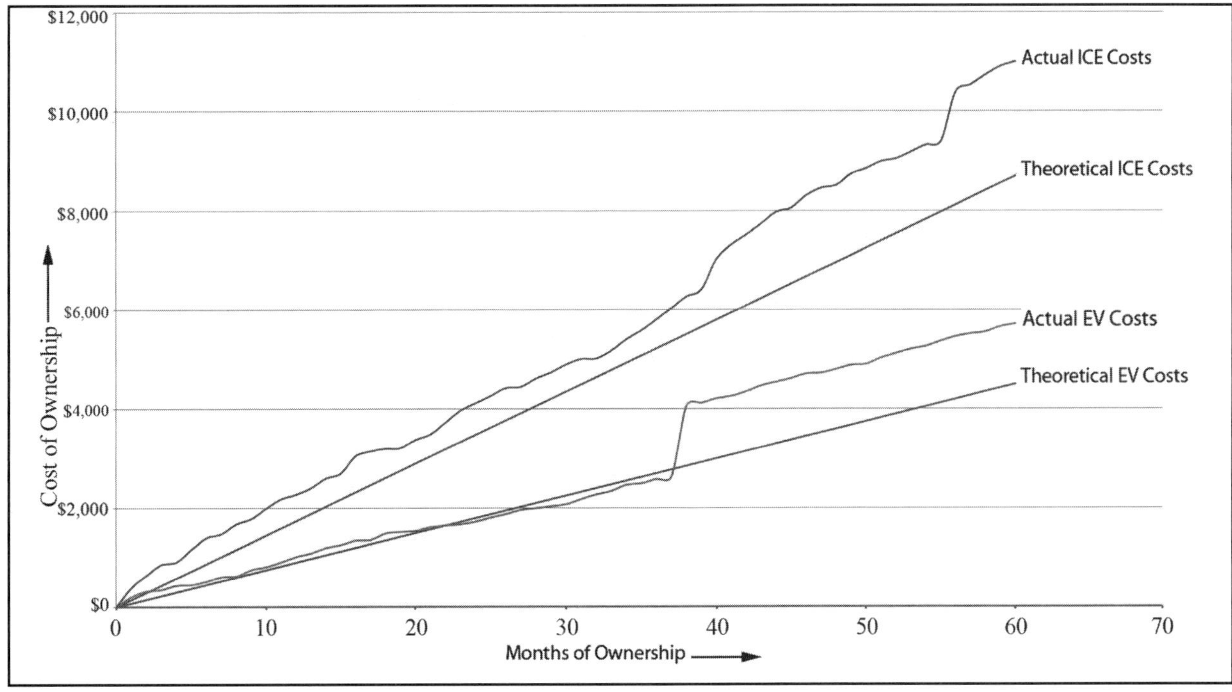

The theoretical operating costs (fuel, maintenance, and repairs) of an electric vehicle are lower than those of a conventional ICE-powered vehicle, but does theory align with the real world? This chart shows both the theoretical operating costs, as well as actual measured results for two similar vehicles put into service at the same time in late 2000. Both cars averaged 12k miles per year. The bottom plot is a1995 Honda Civic that the owner converted to electric power in 2000. The large jump at 3.5 years in is the cost of replacing the 120 volt flooded lead-acid battery set. The upper plot shown is a 2000 Honda Civic that was purchased new and averaged 22.8 mpg. The car underwent normal maintenance and repairs, such as oil changes, fuel injector servicing, and radiator flushing. (Note that tire replacement, brakes, and other similar items are not included in these charts, as they are common to both vehicles. Also, the original cost of conversion is not included for the EV, nor is the original purchase costs for either vehicle.)

understand how much electricity costs when supplied by your local utility company. Electrical rates vary widely by region and, often, by the time of day the electricity is used. According to the Department of Energy (DOE), however, the average retail price of electricity in 2009 in the U.S. was 11.96 cents per kilowatt-hour, or $0.12/kWhr. Some areas in the U.S. are higher than this amount, while some are lower, but this is a good average number to use for calculations. During off-peak times (such as at night) electric rates are often much lower than this amount, sometimes by as much as 25% or more. For our calculations here, however, we will use the average worst-case $0.12/kWhr figure.

Next we need to determine how much electricity is consumed by an EV when in operation. The bad news is that every car is somewhat different than the next in this regard. The overall vehicle weight, aerodynamics, system voltage and capacity, tire pressures and condition, operating speed, driving conditions, and how the driver operates the vehicle can each have a big effect on the efficiency of a BEV. Important also is the type of roads and terrain that the EV is used on. For instance, if a BEV is driven moderately on flat, level ground, the overall range will be much longer than if the BEV is driven hard on hilly roads, where disproportionally more electricity is required.

Okay, all that said, an average EV uses between 0.25 and 0.5 kWhr per mile in average driving conditions. For the purposes of our sample calculation here, we will pick 0.4 kWhr/mile as a typical BEV energy usage rate. (Other types of vehicles, such as motorcycles, can achieve performance much better than this figure, while heavy electric vans and trucks are usually worse. I encourage you to peruse the thousands of electric vehicles shown on evalbums.com to see the energy usage rates of other vehicles comparable to the one he or she is contemplating building.)

If our average BEV is driven 50 miles in one day, the amount of energy consumed would be 50 miles x 0.4 kWhr/mile = 20 kWhrs. Now, if we multiply this by $0.12/kWhr, the cost of commuting the 50 miles is equal to 20 kWhr x $0.12 /kWhr = $2.40, or a little under 5 cents per mile.

So how does this compare to an average ICE-

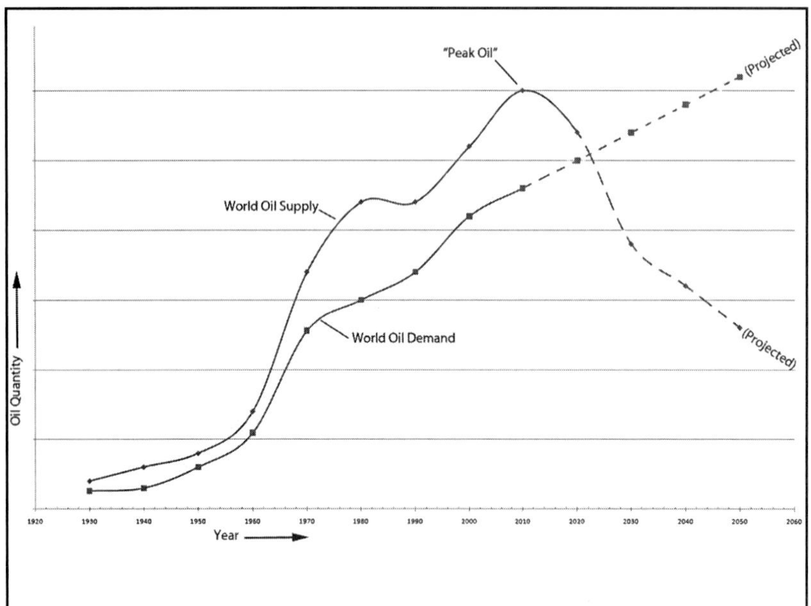

The reason that gasoline costs are predicted to continue increasing is based on the principle of "peak oil." There is a limited supply of crude oil buried under the earth, and many geologists believe that we've reached the "peak" of practical oil supply production. Unfortunately, man's usage of oil shows no sign of slowing down.

powered commuter vehicle? According to the National Highway Traffic Safety Administration (NHTSA), the average gas mileage for new vehicles sold in the U.S. is 24.7 mpg. For our 50-mile commute, this means that we need to burn 50 miles ÷ 24.7 mpg = a little more than 2 gallons of gasoline. The 2009 average price for regular unleaded fuel in the United States is $2.75 per gallon. This means our 50-mile commute costs $2.75/gallon x 2 gallons = $5.50, or 11 cents per mile. In other words, it costs more than twice as much per mile to drive an ICE-powered vehicle then it does a BEV. If we drive an average of 12,000 miles per year, this equates to an annual savings of over $700. And, if we factor in how much gasoline is expected to rise in price over the coming years, the annual savings increase is even more.

Now, critics will often point out that batteries don't last forever and will need to be replaced at periodic intervals. This is correct. If treated well, the average lead-acid battery used in an electric vehicle can only be discharged and then recharged approximately 750 times before it needs to be replaced. For the typical commuter vehicle, this means that every three to five years the battery bank has to be replaced. Given 2009 prices for popular lead-acid batteries, this means that every four years or so a BEV owner must fork out roughly $1200 or more dollars to replace the units.

In addition, the brushes on brush-type electric motors require replacement every 50,000 miles or so. While this falls outside of our four-year window, let's go ahead and include this into the maintenance costs of a BEV. Brush replacement is very easy to perform yourself, so we will only include the cost of the brushes, which run in the neighborhood of $100. So, the annualized maintenance costs of a BEV are roughly ($1200 + $100) ÷ 4 years = $325 per year.

While this sounds like a lot, let's look at the amount of money required to maintain a similar, traditional ICE-powered vehicle for that same four year period. (Note that we will not consider the price of tires, brakes, or other items that are common to both ICE- and electric-powered vehicles, as these costs are the roughly the same.)

Most ICE-powered vehicles require oil and filter changes every 3000 to 5000 miles. In the U.S., the average cost of this service is $35, which is equivalent to $350 in four years of average driving. Next is the cost of coolant flush and replacement, which most manufacturers recommend take place every 20,000 miles or two years. The cost of such a service averages $97 in the U.S. So, for four years of operation, this is equivalent to $194. We also have to calculate the cost of periodic replacement of the air filter, fuel filter, spark plugs and wires, radiator hoses, fan and accessory belts, water pumps, timing belt and other equipment and services that internal combustion engines require, such as the positive crankcase ventilation (PCV) valve, fuel injectors, throttle body maintenance, check-engine diagnostics, and general tune-ups performed on a regular basis.

Note that this list doesn't yet include things like oxygen sensors, exhaust system repairs, muffler and catalytic converter replacements, and the countless other items, components, and subsystems that require repair and replacement on internal combustion engines. The U.S. Department of Transportation estimates that the 2009 cost of these repair and maintenance items for the average passenger vehicle or light duty truck is $420 per year, or about $420 – $325 = $95 more per year than the BEV.

But we're not finished counting the cost savings yet. Owners of BEVs can often take advantage of federal and state tax deductions and credits for owning a green vehicle. For example, at the time this book was written, the U.S. government was offering a direct tax credit for electric vehicles and light trucks of up to $7500, depending on battery capacity. Many states offer similar incentives, as do some cities. Good luck trying to get that type of

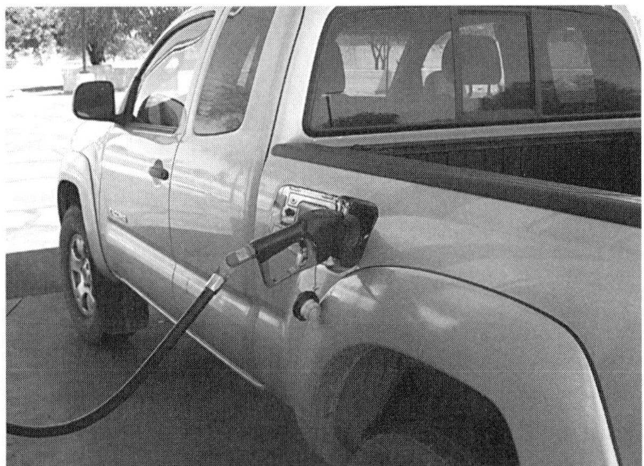

One of the benefits of owning an EV is never having to stop at a filling station for gasoline or diesel again. Charge up over night while you sleep, and you will have a "full tank" of electrons waiting for you in the morning.

The myth that there are few places to recharge an EV is just that: a myth. Any home or business that has electric power can serve as a charging station for an EV. Further, many forward-thinking business establishments, such as this used bookstore in Tucson, Arizona, have begun providing free recharging for their customers. Not only is this approach good for the environment, but it's good for business, too.

Typical battery recharging times for most EVs range from 6 to 10 hours, depending on the battery type, the depth that the batteries have been discharged, and the input voltage to the charger. Courtesy Electric Car Company of Utah.

benefit with your ICE-powered vehicle.

Saving Time with an Electric Vehicle—Imagine never having to stop on your way to work to fill up the gas tank in your car. Owners of battery electric vehicles tend to smile whenever they drive past a filling station, knowing that they never have to take the time to pull in, wait for an open pump, get out and pay for the fuel, pump the fuel, and then negotiate their way back out of the filling station lot and onto the road or highway. Because they've been charged overnight in their garages, BEVs leave their homes fully charged up. Yes, it takes hours to recharge a depleted bank of lead-acid batteries, but most people take care of this overnight as they sleep. Recharging is extremely simple: drive home, plug the charger into the car, go into the house. A quality charger is designed to shut off automatically when the battery is topped off. No hassles, no fuss, no delays in the morning when you're running late for work or trying to get the kids to school. Just unplug the car and drive away. What could be faster?

Owners of BEVs also spend significantly less time in vehicle repair shops. Imagine never having to sit in another car repair shop's waiting room, reading year-old magazines and drinking stale coffee while the mechanics change your oil and filter. All of the maintenance and repair items we talked about earlier in the section on economic advantages also have time-saving benefits. There are no air filters, fuel filters, spark plugs and wires, timing belts, or PCV valves to replace. There are no fuel injectors that need servicing, or radiators that need flushing and refilling. There are no tune-ups to perform, no mufflers to repair, no catalytic converters to replace. Image how much time you could save in this one area of your life if you owned a BEV.

And, as the late night infomercials say, "But wait, there's more!" For instance, if you own a BEV in a town or city that requires annual or biannual emissions testing, you are exempt. Per the Environmental Protection Agency, the average amount of time spent driving to, waiting in line for, getting tested, and then driving home for an emissions test is 2.5 hours. Imagine not ever having to waste a lunch hour or weekend afternoon again getting your car tested.

A Hundred Other Reasons—Okay, so maybe there aren't a hundred additional benefits of EV

A typical charging station that can be used by the public.

ownership, but there are quite a few additional advantages to owning an electric vehicle. For instance, because such a large fraction of the electricity produced in the U.S. is made from coal, hydroelectric, wind, and solar, owning a battery electric vehicle actually reduces our country's dependence on foreign oil. In other words, owning a BEV is a patriotic act that helps keep money out of the hands of the dictators and despots that supply some of the oil we use in the U.S. to create our gasoline. Want to hurt the bad guys who control the flow of oil? Replace your ICE-powered vehicle with a battery electric car or truck.

Battery electric vehicles are also very quiet in operation. In fact, at a stoplight, a BEV creates no noise whatsoever. Imagine how much quieter cities and towns would be if their streets were filled with commuters driving to work in near-silent BEVs instead of ICE-powered cars.

Owning an electric vehicle is also a great way to improve your social life. There are countless clubs and organizations around the country dedicated to electric vehicles, alternative energy, and other green activities and practices. The Electric Automobile Association (EAA) has chapters in every state and most major cities and towns. These groups are filled with enthusiastic and knowledgeable people who want nothing more than to get to know you and talk about electric cars.

There are also many online and Internet-based groups, forums, list-serves, and email digests that you can subscribe to for free. For example, one of the largest online groups is the Electric Vehicle Discussion List (EVDL), which is free to subscribe to. Be warned, however, as it can be very addicting to read, learn, discuss, and debate any and all aspects of electric vehicle ownership with the experienced and friendly enthusiasts that populate these types of groups. The amount of information and experience these people have under their belt is staggering...plus you would be hard-pressed to find a more congenial and friendly group of enthusiasts.

We could go on listing advantages to EV ownership, but by now you understand that there isn't just one reason to consider an electric car; instead there are many reasons that range from the selfish to the altruistic. Save money, protect the environment, or make a statement. Cut time off your weekly commute or meet interesting people. Learn and practice a hands-on hobby. Reduce the country's dependence on foreign oil, and help everyone prepare for a future run primarily on electricity.

For the average commuter, a significant fraction of vehicle time is spent at idling in stop-and go traffic and waiting for stoplights to turn green. When not moving, the motor in an electric vehicle consumes essentially zero energy and emits zero pollution. Compare that to an internal combustion engine that has to keep running during all this time, releasing heat and pollution, and wasting expensive fuel.

Okay, that said, it's time for a warning. The list of advantages are long, and the reasons are compelling but before you run out and jump headfirst into an EV project, it's time to stop and consider the negative aspect of BEV ownership. Yes, there are some downsides to owning and operating an electric vehicle. Let's take a look at them next.

Disadvantages

Now it's time to talk about some of the reasons that a battery electric vehicle may not be a good choice for you to consider owning and operating. Yes, this is a book that promotes the idea of widespread EV use in society, but simply espousing the advantages of something like EVs without discussing the negatives can be a bad thing overall. No one wins if someone is mislead and purchases o builds a vehicle that disappoints them. The best case is they will have wasted their time and money. The worst case is they will broadcast their displeasure to other people and possibly deter them from considering an electric vehicle for themselves. Like most things in life, there is no one-size-fits-all solution to the problem of personal transportation. In this section we will look at reasons you might consider not purchasing or building a battery electric vehicle.

Limited Range—Let's start the list of EV disadvantages with a simple one: range. Battery electric vehicles are not well-suited for long distance travel. The maximum range of a lead-acid battery-equipped BEV is usually around 50 to 60 miles between charges. Lithium-type battery-equipped BEVs can get over 100 miles or more on a charge, but the cost of these types of battery banks is substantial.

For most consumers, 50 miles is more than adequate for their daily commuting needs, but for others it may not be. Estimate your typical daily commute. According to the U.S. Department of Transportation, the average one-way distance that U.S. workers travel to work and school is 16 miles. Only eight percent of workers travel farther than 50 miles, but if you're in this minority group, then you probably should not consider an electric vehicle as your primary means of transportation. A hybrid car, a small turbo-diesel, a motorcycle, or even carpooling in a normal ICE-powered vehicle may be a better choice for you.

One thing to factor in when considering range is whether you can recharge at your destination. For example, some commuters who commute 40 or 50 miles one-way are able to recharge their vehicle batteries during the day while they're working. Then, at the end of the day, their vehicle is ready for the long commute home, where it will be recharged overnight for the trip back in the next morning. If you live 50 miles from your place of employment and you can't recharge like this, an EV may not be an ideal solution for you. In addition, the same principle applies if you need to take many and frequent trips during the day, without leaving adequate time between trips to recharge the batteries. Unfortunately, recharging a battery bank is not as fast as refilling a fuel tank. In fact, it can take 6–10 hours to replenish a depleted battery bank.

Terrain—Terrain is another consideration. If you live in a very hilly or mountainous area, an EV may not be a good choice. It takes significant energy to drive any vehicle, including an EV, up a hill. For a BEV, this extra energy demand can dramatically shorten the operational range of a vehicle. A reduction in range of half or more can occur, depending on the severity and mileage of the hills. This effect is even more pronounced if you live at the top of a long, steep road or highway; batteries are usually at their most inefficient toward the end of their operating range. If you live 40 or more miles from your place of work and have a long uphill climb as you approach your house, an EV may not actually make it home.

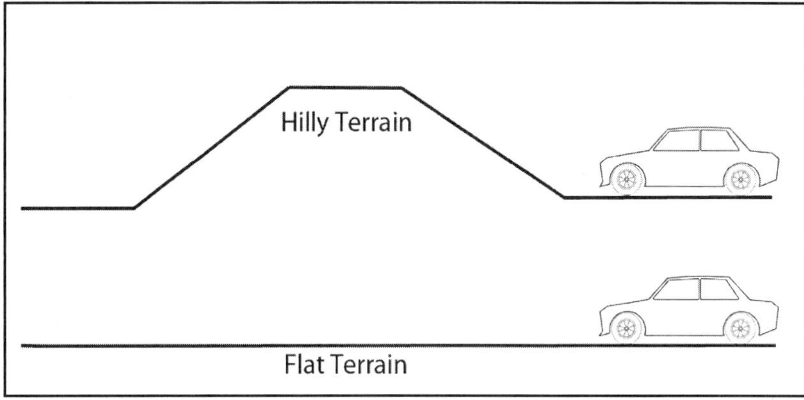

Electric vehicles work very well on relatively flat ground. If you live in a mountainous or hilly area, however, they may not be the best solution for you, as the energy usage to climb a hill is significant and can shorten the usable range of an EV. Higher-end electric vehicles can utilize regenerative braking to help recapture some of the energy associated with braking, but the effect is not huge. If you need to drive up and down large hills on your normal commute, take a good look at what range you can expect to achieve.

Vehicle Preferences—Another thing to consider is whether you want to drive an automatic-transmission equipped vehicle. If the answer is yes, then a BEV may not be the right choice for you. While some enterprising individuals have successfully built automatic EVs, the vast majority of vehicles are equipped with manual transmissions. The reason is one of efficiency; automatic transmissions are often significantly less efficient than manuals, which means that more energy is required to operate them.

Along the same lines, you need to consider the type of vehicle you want to drive. Small, efficient economy cars and compact trucks make the best conversions. Large, heavy sedans and SUVs don't make for good conversion projects. Neither do vans and boxy vehicles, or ones fitted with a lot of power-hungry and heavy accessories, such as air-conditioning (though it is possible to install A/C in an EV). Again, the reasons boil down to simply maximizing efficiency. The heavier and more inefficient the vehicle is, the more batteries required to get reasonable range. But more batteries mean heavier weight still, which subtracts further from the range.

Finally, an electric vehicle is not always the best choice if it is going to be the only car you own. Because of their limited range, EVs are not the ideal one-car solution for a family. Instead, many people own an EV as their second car, using it primarily for in-town commuting, shopping, and school duties. The ICE vehicle is used for weekend trips and longer distance commutes that are more infrequent in nature. That said, it is possible to have an EV as your one and only car; if you do go this

It's possible to convert an automatic transmission vehicle to electric power, but it's probably not going to be easy. Most EV experts suggest sticking with a manual transmission, as it will be much easier to convert. That's the bad news. The good news is that in most driving situations, an EV equipped with a manual transmission is extremely easy to drive. In fact, many EV drivers find that they can simply leave their car in second or third gear for most of their in-town driving. Because electric motors don't stall in the traditional sense, you can leave your foot off the clutch and literally slow the vehicle down to zero miles per hour with no harmful effects.

EVs are supposed to be slow and stodgy, right? Wrong! Try telling the owner of the amazing KillaCycle drag bike that his electric vehicle can't keep up. The motorbike has 500 horsepower, can accelerate from zero to sixty miles per hour in less than one second, and covers the standing quarter mile in less than eight seconds flat. Courtesy KillaCycle.com.

route, you can always rent a traditional ICE-powered vehicle for the occasional highway trip.

Electric Vehicle Myths

Beyond the simple advantages and disadvantages of electric vehicle ownership are a number of myths, urban legends, and downright untruths that you need to understand. Let's look at—and debunk—a few of the most common.

The Range of an Electric Vehicle Is Too Low—Whenever this statement is made by someone, the best thing to do is to respond with a question: "What range is too low?" As we mentioned earlier, the average daily one-way commute of U.S. drivers is 16 miles, or 32 miles round trip. Most basic BEV conversion vehicles can achieve 40–50 miles of range between charging with little difficulty. If you factor in the ability to "opportunity charge" while at work or school, the average BEV can fairly easily achieve nearly 100 miles of daily range. Still not enough? Consider high-end lithium-type batteries, which can add significantly to this figure (but they're not cheap). Or, if you're buying (i.e., not converting) your electric vehicle, you probably can do even better. For instance, the high-performance Tesla Roadster can achieve nearly 250 miles between charges.

Electric Vehicles Are Too Slow—Again, the way to address this misstatement is to ask, "How slow is too slow?" Electric motors develop full motor torque at zero rpm. This means that EVs accelerate extremely well if they have a properly sized motor and controller. Electric vehicles also can achieve illegally high top speeds with little fanfare. The Tesla Roadster, for example, takes only 3.9 seconds to go from zero to sixty miles per hour and is electronically limited to a top speed of 125mph. Inexpensive home-built BEVs usually don't perform at the Tesla's Porsche-busting specifications, but they do pretty well in their own regard. It's very easy to convert an ICE-powered vehicle to electric power and not have any significant degradation in either acceleration or top speed. Batteries may be heavy, but modern electric motors are very efficient and powerful. Electric vehicles are simply not slow.

Electric Vehicles Take Too Long to Recharge—By now you know the question to ask is: "How long is too long?" Yes, it takes time to recharge a battery bank, but so what? If you are using a BEV the way they're supposed to be used, you should never have a problem in this regard. Drive to work in the morning and return in the evening. Plug the vehicle into an outlet in your garage. Wake up the next morning, unplug the car, and drive to work again with a full battery bank. What could be simpler?

The average time it takes to recharge a battery bank depends on many factors. How depleted is the battery? What type of battery is it? What is the voltage and current capability of the household electricity used to charge the system? What type of charger are you using? All of these things have an effect on charge time, but if we just look at the

What EV Owners Say...

In a recent poll of EV owners, the most important benefits stated were that they are cleaner, quieter, less expensive to operate, more reliable, and easier to work on than ICE vehicles. Owners also report that one of the nicest side benefits to EV ownership is that you get to refuel in your garage instead of at a gas station. The convenience factor is important, but for some people, not supporting the oil industry and helping to reduce dependence on foreign oil is the most important reason of owning an EV.

average EV, a typical recharge time is on the order of 6–10 hours or so. Hopefully, you're getting this much sleep every night!

Electric Vehicles Won't Work in Cold (or Hot) Weather—This is patently false. While most batteries do suffer some performance degradation in very low (and/or very hot) weather, the effect is usually small. Range is the biggest item affected by cold weather, but this means that you simply have to install slightly more battery capacity when planning a conversion. There are literally thousands of BEVs driven in severe cold weather settings. The same is true for hot climates; there are thousands of BEVs operated daily in hot deserts.

You Can't Have Air Conditioning (or Power Windows, Steering, etc.)—This statement is also false. There are many BEVs that have all the creature comforts that traditional ICE-powered vehicles do. Air conditioning, cabin heat, power windows, power brakes, power steering... all are possible in most EV conversion projects.

Electric Vehicles Are Maintenance-Free—This one is almost true, but not quite. All vehicles require some level of maintenance, from periodic replacement of things like tires and brakes, to lubricating the suspension and chassis, to adding windshield wiper fluid. Unlike an internal combustion engine-powered vehicle that requires oil changes, coolant flushes, air filters, and other tune-up needs, electric vehicles require very little additional maintenance. They are not, however, completely free from occasional servicing. For instance, if your BEV uses traditional flooded lead-acid batteries, you will need to periodically check the state of the electrolyte, and possibly add water to the cells of the batteries. You may also need to change motor brushes every five years or so. The batteries themselves will also likely need to be removed and replaced every 3-5 years. Other than those, there are very few things that an electric vehicle requires in terms of maintenance. Compared to internal combustion engines that have literally hundreds of moving parts and complex coolant, fueling, induction, exhaust, and lubrication needs, an electric motor has just a few moving parts, and the Mean Time Between Failure (MTBF) of these parts is considerably longer than most ICE parts.

You Can't Register (and/or Insure) an Electric Vehicle—Tell this to the tens of thousands of BEV owners currently operating their vehicles today in the U.S. Registering an electric vehicle is usually quite simple. In fact, once you get a waiver for emissions testing, you will never waste another minute waiting in line at the local inspection station. Insurance is also relatively easy to get, though you may have to show your agent that the EV component installation was done in a safe and professional manner. You also might want to consider insuring your car for more than you would if it were an ICE; electric vehicles often attract attention from curious bystanders.

Electric Vehicles Are Expensive to Own and/or Operate—This is another blatant untruth frequently told by people with vested interests in keeping the petroleum industry afloat. As we demonstrated earlier in this chapter, simple mathematical calculations show that a typical electric vehicle is cheaper to operate and maintain on an annual basis than an equivalent ICE-powered vehicle.

Information Is Key to a Successful EV Build

Most maintenance and repair issues with EVs occur because they are hand-built by inexperienced and/or uninformed builders. These people make mistakes because they don't know any better. Incorrect or low-quality parts get used, and those tend to fail early. Things get put together wrong, too, and then fail. Safety and backup systems also get left out, so the driver can (and does) break things. Parts get pushed to their absolute maximum ratings, rather than being conservatively rated. Things get left out to save money or time, or because the builder doesn't know any better. The secret to avoiding these problems is to do it right the first time—and that starts with knowledge and understanding.

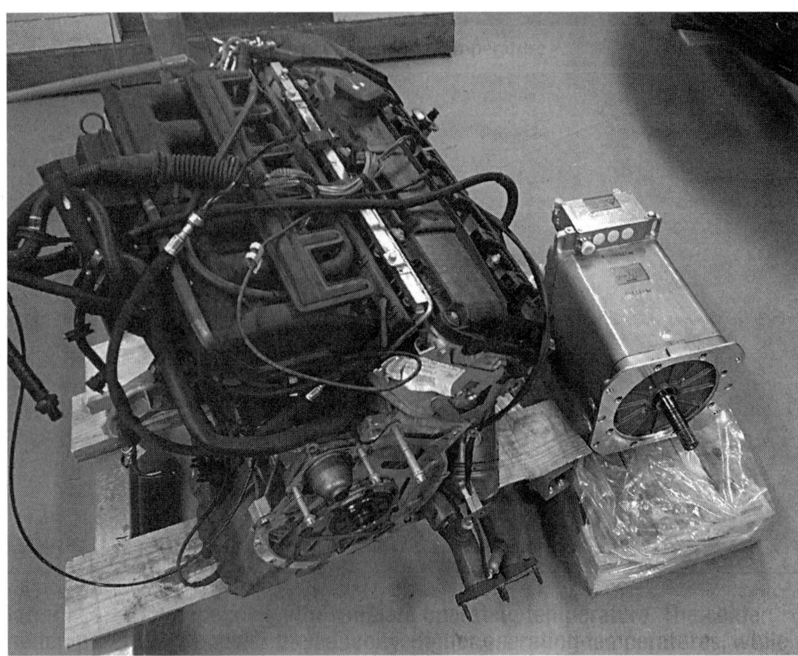

There is a misconception among some people that electric vehicles are complicated. Nothing could be further from the truth. Shown here is a comparison of an internal combustion engine (ICE) that was removed from a BMW, with the electric motor that replaced it. Factor in all the other supporting equipment needed for the ICE, such as radiator, oil cooler, exhaust system, pollution controls, starter circuitry, fuel tank, pump, injectors...well, you get the idea. Electric vehicles are much simpler in comparison. Courtesy Bob Simpson.

Electric Car Batteries Are Problematic and/or Wear Out Very Fast—Lead-acid batteries are used in the majority of EV conversions. These units have been around in one guise or another for nearly a century, and the technology has steadily and incrementally improved every year since their inception. The result is that the modern variants of the lead-acid battery are now very stable, reliable, and can deliver hundreds, if not thousands, of charge-recharge cycles if they're treated well. In a well-maintained application, a set of lead-acid batteries can last up to five years or more before requiring replacement. Lithium-based batteries are even more impressive, with much longer lives and with much larger allowable numbers of charge-recharge cycles.

Lead-Acid Batteries Are Bad for the Environment—Yes, lead is a toxic substance that can harm both humans and the environment, and lead-acid batteries are literally filled with lead. But so what? The lead is contained and not released into the atmosphere under normal use. And, when it's time to replace the batteries every four to five years or so, the old batteries are extremely recyclable. In fact, lead-acid batteries are one of the most recyclable products used by modern man. More than 98% of the materials used in a battery can be recycled and reused with no harmful effects on the environment. Compare that to how much damage and pollution occurs when one gallon of gasoline is consumed in an ICE.

Electric Vehicles Aren't Safe—Built correctly, an EV is every bit as safe as an ICE-powered vehicle. The basic chassis and body is built to the same standards, crash test standards are no different, and the same personnel safety equipment is used (e.g., seat belts, air bags). If an EV is built and maintained properly, its batteries won't blow up in accidents and the owner won't be electrocuted. Let's say this another way: Your typical ICE-powered vehicle has 15–20 gallons of highly flammable liquid sloshing around in a tank that's bolted underneath the body. Every 200 miles or so of use, the owner has to physically refill this highly toxic and explosive liquid. Do you feel safe performing this task? Recharging a battery bank is eminently safer.

You're Out of Luck If Your Batteries Run Out of Electricity—Well, yes, but you are also out of luck if your ICE-powered vehicle runs out of gasoline. What's the functional difference? Or, more important, how often do you run out of gas on a regular basis? Not often. The same is true for electric vehicles. Instead of a gas gauge, electric vehicles typically are fitted with voltmeters and/or state of charge instruments that provide a measure of how much energy remains in the battery bank. And if you design your system correctly from the start and recharge at night while you sleep, you will probably never run out of electricity on a normal commute.

If you do manage to somehow screw up and run out of battery energy, recharging is just as close as the nearest 110-volt household wall outlet. In other words, just about any house, store, workplace, school, or other modern building has the ability to emergency recharge an electric vehicle. Try doing that with your gasoline-powered car or truck!

Electric Vehicles Pollute Just as Much as ICE-Powered Vehicles—While it is true that pollution is created at many coal- and oil-powered electricity plants, the equivalent amount per mile of vehicle use is still much lower than that of an ICE-powered vehicle. And let's not forget that more and more renewable energy sources are being built every year in the U.S. Wind, solar, and hydroelectric power plants comprise a growing segment of our energy supply needs. As these systems take hold, the amount of pollution released by power plants providing energy to charge BEVs will just continue to get lower and lower.

The Country Will Need to Build More Power Plants—There's currently enough off-peak electricity in the U.S. to power roughly 80% of all U.S. vehicle demands if everyone drove an EV. In other words, if four out of every five drivers sold their ICE-powered car or truck and switched to electric power overnight, we'd have enough electricity to cover their needs.

You Can Mount a Wind Turbine on the Roof of Your Car and Charge Up the Batteries While You Drive—While you could conceivably mount a wind turbine on the roof, and it could conceivably charge the batteries while you drove, the amount of energy it would take to overcome the drag and spin the turbine would be more than you could recapture. In other words, perpetual motion still can't be achieved by us mortals.

You Can Use Solar Panels on the Roof of Your EV to Charge the Batteries—Solar panels can be used to recharge an EV's battery bank, but it is usually not cost effective to do so, and this assumes you use large, fixed-mount batteries at your home. It takes a relatively large surface area of photovoltaic panels to provide enough power to recharge a BEV. Unless you drive a stretch limo, there probably isn't enough area on the roof of your car to mount the required number of panels.

Thoughts on Pollution...

On the subject of pollution, EV expert Lee Hart has this to say, "Even with their pollution controls, cars are incredibly dirty. Think of it this way: Close your garage door, and run your car inside for an hour. The amount of pollution it would produce would make you sick, or could even kill you. In contrast, you could leave your EV running in your closed garage for days with no significant effect on the air quality."

Calculating Energy Usage

A rule of thumb to calculate the approximate energy usage rate (in kWhrs/mile) for a normal BEV is:

Vehicle weight ÷ 10,000

For example, for a 3000-lb vehicle the usage rate will be:

3000 ÷ 10,000 = 0.3 Kwh/mile

Chapter 3

The EV Conversion Process

"You can't always get what you want; but if you try sometimes you might find you get what you need."
—The Rolling Stones

Converting a vehicle to electric power doesn't have to be expensive. Lightweight, simple, and older vehicles, such as Volkswagen's venerable Bug have been converted by countless enthusiasts for relatively low cost. A small DC motor, a simple controller, and a modest set of flooded lead-acid batteries are the building blocks of economical EVs. This particular EV uses an 8" Advanced DC motor, Curtis controller, and a 96-volt battery bank.

Buy vs. Build/Convert

So, now you understand how a battery electric vehicle works. You also now appreciate the pros and the cons of EV ownership, plus you can tell the difference between the facts and the myths surrounding them. You've looked at a couple dozen Internet sites, homepages, and blogs that detail various individuals' car and truck projects. You're ready to take the leap into EV ownership, but there's a big question you haven't yet addressed: Should you actually build an electric vehicle, or should you instead simply buy one that is already built and running?

Converting an ICE-powered vehicle to electric power can be a very rewarding project to undertake, but it's not for everyone. Building an EV can be either vexing or challenging, depending upon your patience, abilities, and personality. There are also costs to consider; depending on the type of car and electrical system you choose, converting a vehicle can be either cheaper than, or more expensive than just purchasing a pre-built conversion that someone else has already created. A new conversion project can and will require significant time and resources, and, frankly, many people are better off just buying a new or used EV than tackling a ground-up conversion by themselves.

Now, don't get the wrong message from this warning. This book is clearly focused on converting ICE-powered vehicles to electric power. With the act of purchasing this book, you have shown that you are someone interested in the idea of conversion, but it's vitally important that you enter into an EV project with your eyes wide open. It is not all sunshine and roses, as they say, to build an electric vehicle. It costs money, it takes time, and it can be frustrating. You need some basic mechanical and electrical wiring skills. You need to be persistent, even somewhat stubborn. You cannot take shortcuts, and you absolutely have to know what you're doing every step of the way. Do you have these traits? Are you willing to invest a significant amount of time and money converting a car or truck?

If the answer is yes, then great! Building an electric vehicle can be both a worthwhile and satisfying endeavor. But where does one begin? At first, the process of conversion can seem overwhelming. Where should you start? What needs to be considered? How do you proceed? There are many different approaches, but a six-step process seems to work well:

1. Define Your Requirements and Abilities
2. Define Your Budget
3. Select a Vehicle
4. Select the EV Components and Parts

Planning

When planning your first EV project, you might consider copying what someone else has already done that closely fits your needs. Benefit from their experience. If you don't have to reinvent the wheel, why do it?

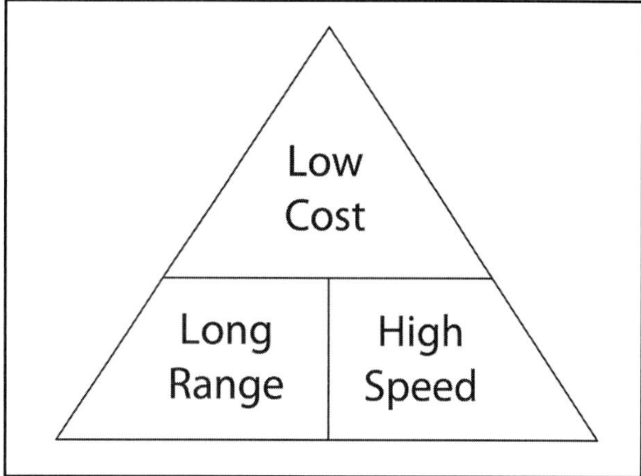

There is an old joke among experienced EV builders that says there are three things people usually look for in an EV: speed, range, and low cost. Choose any two. If you want range and speed it will cost a lot. If you want low cost, you can usually get speed or range but not both.

5. Locate Work Space and Tools
6. Convert the Vehicle and Drive

Let's look at each one of these steps in a little more detail, starting with defining requirements.

1. Define Your Requirements and Abilities

The variety of different types of electric vehicles you can build is vast. From a simple electric bicycle or go-kart to an advanced high-powered specialty sports car, the possibilities are nearly limitless. To help narrow down your choices, it is suggested that that you begin with a list of your vehicle requirements. Once you've specified what you want an electric vehicle to do, choosing the type of vehicle and selecting the system components will be made much simpler.

For instance, how many people will the vehicle need to carry? Just you? You and one passenger? An entire family? You may want a 2-seat sports car, but you may need a sedan or van.

You should also ask yourself how far do you need to travel on a single charge? A few miles to and from work on city streets? Fifty or more miles on the highway? Range costs money, and it's not necessarily wise to design and build an EV capable of traveling a hundred miles between charges if most of the time the car is used for short in-town hops.

Similarly, how fast will it need to go? Will you be driving the car in a town or city? On rural roads? On the highway? Any vehicle, whether it's electric or gasoline powered, needs to be able to keep up safely with traffic.

You should also consider what features and creature comforts you need. For instance, do you need air-conditioning where you live? What about heat? What power accessories do you need? Power windows? Electric seats? Convertible top?

Another consideration is payload capacity. Do you need the vehicle just to transport yourself to and from the grocery store, or do you need it to carry heavy and/or over-sized loads from, say, a hardware store?

Also, where will the vehicle be charged? Only at home? At your destination, too? How far will it have to go between charges? How much time will you have to charge the vehicle? Do you have any special needs in a car, such as an automatic transmission? High ground clearance? The ability to drive off road?

You also need to ask yourself if you are considering a conversion because you need a means of transportation or whether you're looking for a technical challenge to undertake. Do you want the vehicle to stand out from the crowd or do you want it to blend in and not attract attention. Ask yourself why you want an EV? Status, economy, or social concerns? Will this be your only vehicle? A back-up car? Your primary means of transportation? The answers to these questions will help focus your needs.

You also need to assess your own mechanical abilities and skills. Are you handy with basic hand tools? What about more advanced equipment like grinders, welders, and soldering irons? Do you have access to these types of tools? Would you be comfortable fabricating your own battery boxes or motor mounts? Or would you be more comfortable simply bolting together the components from a conversion kit?

Time is another important consideration, as is the work space where you intend to perform the conversion. Do you have a place to work on the car that is safe, secure, and well-lit? How much time do you have to convert the vehicle? Will it just be you working on the conversion, or will you have help? How much of the effort will you need to contract out to the experts?

The list of questions you need to ask yourself is long, but even more important are the answers you provide. You need to be honest with yourself and your abilities. The old adage rings true when it comes to electric vehicle conversion projects: You can't get what you need until you define what it is you need. There is also the issue of needs versus

If you're contemplating your first electric vehicle conversion project, it's best if you keep the project simple. Bicycle conversions make excellent introductory conversion projects. These can range from the ultra-cheap, cordless-drill-powered bikes, to high-end, lithium battery-based two-wheelers. Most electric bikes fall somewhere in the middle, with the vast majority constructed from kits that are available for purchase online. Manufacturers like Amped Bikes make excellent, affordable, and very easy to install kits. And with low-priced lithium-based batteries available online, there is no reason you can't construct a long-range, highly reliable e-bike for relatively little expense. The author's own electric, full-suspension electric bicycle is shown here. This converted mountain bike routinely travels 30+ miles on a single charge and can top 30mph on level ground.

wants. You may want an electric sports car, but you may need a 4-passenger vehicle. Sure, we all want 0-60 in 4 seconds, a 200-mile range, two-hour charging, air-conditioning, an automatic transmission, and power seats, but some or all of these things may not be practical or cost-effective for you. Planning an EV conversion project almost always boils down to compromises, which is made simpler when you begin with a determination of your requirements and abilities.

The Chevrolet S10 pickup truck is another very popular vehicle to convert to electric power. The base vehicle can usually be purchased fairly inexpensively, and it was designed from the factory to carry heavy loads (such as those imparted by a large array of lead-acid batteries). A number of suppliers sell components and conversion kits for this vehicle. Other popular trucks to consider for conversion include Ford Rangers, Nissan, Toyota, and Mazda B-series mini trucks. The S10 shown here has been converted to electric power using an AC motor and controller and a high-voltage battery system.

2. Define Your Budget

The next step after determining your vehicle requirements is to define your budget for the conversion. There is a wide range of prices that can be expected during construction, so determining your budget will help focus the project and aid in component selection. An electric vehicle conversion project can be carried out for relatively low cost, but undoubtedly trade-offs and compromises need to be made. Of course, if money is no object then the sky is the limit, but for the rest of us mere mortals, establishing a budget is an important step to undertake.

It does no good to plan an expensive high-tech electric system (e.g., AC drive and controller, lithium batteries, regenerative braking, etc.) if your budget can only support a low-cost system (e.g., DC drive and controller, lead-acid batteries, no regen, etc.). The vehicle choice will affect this decision, too; large, heavy vehicles will require more battery capacity (read: more money) than smaller, lighter vehicles. The cost for parts and labor to convert a car can range from $3,000 to $5,000 on the low end, to $25,000 or more on the high end. And this does not include the cost of the donor car.

Besides the type of component and level of sophistication of the system you desire, another factor that can greatly affect the cost is how much of the work you will do yourself, versus how much you will need to outsource to professionals. How much do conversion companies charge in your area? Also, how good are you at searching out deals and scrounging parts? If you decide to buy all of your parts and components new, the price will undoubtedly be higher than if you find pre-owned equipment on eBay, craigslist, or other EV classified ads. Auto recyclers and junkyards sometimes have components that you can use.

You might also consider purchasing a running (or even non-running) EV that someone else has previously converted to swap over the equipment to your project vehicle. Being creative and resourceful can greatly reduce the cost of conversion, but it takes time to shop around, and you also have to be sure that you're purchasing equipment in good running condition. Motors, adapter plates, contactors, and similar equipment are generally very reliable and easy to test prior to purchase. Electronics, however, are trickier; purchasing a used controller or DC-to-DC converter is something you should consider only if you're confident that the equipment is in good shape. Further, some items such as batteries are generally recommended to be purchased new, not used.

The bottom line is that all these factors will affect

the cost of your proposed EV conversion. Establishing ahead of time how much you can commit to the project will help focus the answers to the questions. Without knowing the allowable budget, you could head down the wrong (i.e., too expensive) road, and not discover that fact until it is too late. Determine your budget now, before moving ahead to the next section, which is selecting a vehicle. In fact, depending upon the budget you've determined, you might even want to go back to Step 1 and refine your list of needs vs. wants.

3. Select a Vehicle

Once you've decided on your vehicle requirements and general budget, the next step is to select an actual vehicle make and model to convert. The good news is that your list of requirements and budget will greatly aid you in chassis selection. The bad news is that there is no "best" car to convert, even if you have your requirements and budget fully understood. This is because every person has their own notions of what would make a nice vehicle to drive. These differences can be big (e.g., truck versus car), moderate (e.g., Chevrolet vs. Ford), or small (e.g., four-door vs. two-door). In the end, the selection of a conversion vehicle will depend largely upon your personality. What vehicles do you find nice looking? What makes you turn your head when it goes by? What vehicle will you want to show off to your friends and coworkers after conversion? If you build a reliable, well-running EV, you will likely own it for years, if not decades. Choose a vehicle that makes you happy every time you sit in it to drive somewhere.

Lighter Is Best—Okay, all that said, there are a few additional things to consider beyond personal aesthetics when choosing between different donor platforms. For instance, lightweight vehicles are generally better than heavy vehicles to convert. Colin Chapman, the founder of the legendary Lotus Cars Company, is often credited with the famous design philosophy: "To go fast, add lightness." Electric vehicle constructors have a slightly different take on the saying: "To go far, add lightness." In other words, when given a choice between two similar vehicles, the better choice is usually the one that weighs less overall. Heavy vehicles require larger motors and bigger battery banks to achieve the same performance as a lighter vehicle. Said another way, if the same battery and motor combination are installed in two vehicles, and one weighs just five percent more than the other, the overall range and speed will be reduced in the heavier car. There is a saying among knowledgeable EV constructors that sums this point up well: "Strive for less iron (vehicle weight) and more lead (batteries)."

Small sedans, such as the Geo Metro, make excellent EV conversion projects. A simple DC motor and controller, along with an 84-96 volt battery bank, can provide relatively long range and peppy performance for an inexpensive conversion price. Other similar vehicles that fall into this category include Dodge Neons, various Fiats, Ford Festivas and Fiestas, Honda CRXs and del Sols, Hyundai Accents, Mazda 121 and 323s, MINIs, Nissan Sentras, and Toyota Corollas, Tercels, and the Yaris.

Consider GVWR—While choosing a lightweight donor vehicle to convert is important, so is selecting a vehicle that has the ability to carry the extra loads imparted by the heavy battery bank it will have to carry. The key to addressing this is by way of the vehicle's GVWR number. All modern cars and trucks sold in the U.S. have a Gross Vehicle Weight Rating, or GVWR. The GVWR is the total weight of the vehicle plus the maximum load it can carry. In other words, the GVWR gives an indication of the safe design load that the vehicle's chassis, brakes, steering, and wheels can withstand. When evaluating a possible conversion vehicle for your EV project, look at this figure and determine whether the car can carry the expected load of the batteries. Most U.S. cars and trucks have a GVWR sticker or placard either in the driver's side door or door frame, or under the hood.

When considering the GVWR, you should also consider how much fore-aft imbalance you will likely end up with after conversion. If the vehicle's overall GVWR is sufficient for the expected battery load, but you're forced to put all the heavy batteries at one end of the vehicle, say over the rear axle, this may result in an unsafe condition. If the front-to-rear weight ratio prior to conversion is, 60 percent front and 40 percent rear, you should strive to maintain this approximate balance with the placement of the batteries.

Consider Aerodynamics—Another consideration when selecting a donor vehicle is its aerodynamics. Engineers quantify how sleek, or aerodynamic a vehicle's shape is by way of something called its coefficient of drag, or Cd. A large, boxy vehicle will

Medium-sized sedans and hatchbacks are very popular as first-time conversion projects. In addition to the electric Honda Civic shown here, other popular mid-sized vehicles that you might consider converting include Acura Integras, various Audis, BMW 3-series, Nissan 240SX and Camrys, Ford Escorts & Probes, Mazda 626s & Protégés, Saab 900s, Saturn SCs and SLs, Toyota Camrys, and various late-model Volvos.

Application	Voltage	Amps
Bicycle:	24-48	50
Go-Kart:	36-48	300
Motorcycle:	48-72	400
Low-cost Car Conversion:	72-108	400
Average Car Conversion:	120-144	500
High-Performance Car Conversion:	156-192	600-1000
Race Vehicle:	200+	1000+
Typical OEM Production EV:	312	400

One of the first questions that you need to address in the design process is that of the overall system voltage for your vehicle. Low-voltage systems are inexpensive, but also come with generally low performance characteristics. Conversely, high voltage systems can provide great performance, but they can also be quite expensive to implement. This chart gives typical system voltages for a range of different applications. Also shown are normal amperage draws for each vehicle type. Keep in mind that these numbers are just a guideline, and that your own application may require slightly different values.

generally have higher Cd (i.e., more aerodynamic drag) than a small, sleek vehicle. This factor is important if you intend to drive your vehicle at substantial speeds on highways. The wind drag on a vehicle increases with the square of speed; doubling the speed of a vehicle from, say 30 mph to 60 mph increases the aero drag by a factor of four. Worse, the power required to push the vehicle through the air increases by the cube of the speed; a vehicle at 60 mph takes eight times the electric power to maintain that speed than it does at 30 mph. If your vehicle will be used primarily in town and at lower speeds (i.e., 40 mph or less), aerodynamics may not be a major consideration for you. At higher speeds, however, aero drag can have a significant effect on the maximum speed and range possible.

Price is another factor to consider. It's often not recommended that you convert a car that your neighbor or friend gives you simply because it's free. Just because it didn't cost anything up front doesn't mean it won't cost a lot to get back in roadworthy condition. You should take a long, hard look at any free or low-cost vehicle and compare how much money it will take to make it licensed and as roadworthy as a similar pre-running vehicle. Rust, bald tires, worn-out shocks and brakes, cracked glass, faulty electrical systems...all of these things will need to be repaired or replaced before you pay out the first dime on EV components. Spend your conversion money on the conversion, not a restoration.

The Newer the Better—On that same note, it is usually better to convert a newer vehicle than an older car or truck (up to a point; more on this in a moment). The reasons for this are numerous. Parts and components are generally easier to source for a 2000 Honda Civic than they are for a 1961 Nash Metropolitan. You are also more likely to be able to purchase a factory service manual (FSM) for the newer vehicle than the older one. Similarly, online information for the newer, more common vehicle will be easier to come by. While you won't need the FSM for the internal combustion engine and its subsystems, you will want it available for helping with things like electrical wiring hook-ups (e.g., lights, wipers, etc.), brakes, suspensions, instrumentation, and so on.

The one caveat to purchasing a very new vehicle, however, is one of electronic compatibility. Some modern cars and trucks have electronic computer systems that control not only the ICE, but instruments, lighting, brakes, airbags, and other systems. Make sure that it's reasonably easy to divorce these systems from the control unit so that you can integrate them into the EV electronics.

In a similar vein, it is generally not advisable to convert an automatic transmission-equipped vehicle to electric power. Automatic transmission may have a place in certain specialty EVs, but most electric vehicle conversion projects are better served with manual transmissions. The process of design and construction of EVs is all about maximizing efficiency, lowering weight, reducing drag, improving rolling resistance, and so on. Automatic transmissions are generally 5-10% less efficient at transmitting power from their input shaft to their output shaft than a similarly sized manual. This is simply physics and has to do with how an automatic transmission's torque converter slips at low speed, converting what would otherwise be useful work into waste heat. When consumer

> Perhaps the most important consideration when choosing a vehicle to convert is to select one that you personally like. If you don't, you won't be happy with the results, no matter how well it performs.

vehicles are offered in both auto and manual models, the manual almost always has a higher mpg rating for this very reason. The other reasons to seriously consider a manual transmission for an EV conversion include cost (manual conversion cars are generally cheaper than auto-equipped vehicles), complexity (automatics tend to be more complicated and often require special control electronics to get them to shift automatically), and maintenance (ask any SAE-certified mechanic which transmission type is easier to service and repair; manuals will win hands down every time). In short, manual transmissions are simpler, smaller, lighter, more efficient, and usually more durable than automatics.

Choose a Popular Vehicle—Another major consideration to selecting a vehicle for your first conversion project is to choose a popular car. You may have your heart set on a baby blue 1962 Citroen, and you may well have the skills and abilities to pull it off, but for the rest of us, choosing a car or truck that a number a people have already converted is probably the better choice. Said another way, choosing a popular model means that there will be people available to answer questions and give you guidance on what worked—and what didn't—on their own similar conversion project. Plagiarism may be frowned upon in school, but it's encouraged in the world of EV conversions. It's far easier to copy someone else's designs and solutions than to invent your own from scratch. In addition, some vehicles are so popular that conversion kits can be purchased in one large package deal. There is very little head-scratching and figuring things out with a kit. In fact, some kits can be installed and the vehicle up and running after just 40–60 hours of effort.

Small, lightweight and popular cars for conversion are abundant. Some of the best to consider include the likes of Volkswagen Rabbits and Beetles, Toyota Camrys and Tercels, Nissan Sentras and Maximas, Honda Civics and CRXs, Geo Metros, Ford Escorts and Festivas, Dodge Neons, and Chevrolet Aveos. If sports cars are more to your fancy, there are a number of very good conversion choices that range from older Porsche 914s and MGBs to newer Mazda Miatas and even Corvettes. Trucks are also very popular, primarily because they tend to be inexpensive and are designed from the factory to carry heavy loads like batteries. The most common trucks that are converted are Chevrolet S-10s, Ford Rangers, Mazda B2000s, and Toyota mini trucks. To see what has been converted before, it's recommended that you spend an afternoon or two browsing Internet sites like evalbum.com.

Finally, when considering different vehicles and options, keep in mind the old adage: KISS, or Keep It Simple, Stupid. A simple, popular conversion is infinitely better for a first-timer than a rare vehicle with a complex drivetrain and battery choice. Buy a newer, simpler vehicle, install a basic EV drivetrain and battery bank, and begin driving your car. Save all the bells, whistles, and complexities for your second EV!

High-end conversions can be carried out on any vehicle, but for all-out performance a vehicle like this electric Porsche 911 can offer neck-snapping (and head-turning) thrills. Popular vehicles that fall into this category include Audis, BMW Z3 and Z4s, Chevrolet Corvettes, Datsun Z-cars, DeLoreans, Ford Mustangs, Mazda RX-7s, Nissan 300ZXs, Porsche 911s, 914s, and 944s.

4. Select the Components and Parts

After you have provisionally chosen a vehicle to convert, it is time to begin the process of selecting

Everyone loves a vintage convertible, and silent electric power makes the top-down driving experience even more rewarding. Popular rag-top EV conversions include Alfa Romeos, Austin Healeys, Jensen Healeys, Mazda MX-5 Miatas, MGBs and Midgets, and Triumphs, like the TR6 shown here, which is powered by a WarP 9" motor, Zilla controller, and uses AGM-type lead-acid batteries.

Kit cars are popular for conversion to electric power because they are generally simple in design, lightweight, and designed to be easily assembled and worked on by their owners. Shown here is a sleek fiberglass-bodied 1980 Bradley GT II that is powered by a powerful AC motor and twelve 12-volt gel-type lead-acid batteries. Courtesy Mark Bush.

A fiberglass-bodied roadster that has been converted to electric power. Courtesy Ed Matula and eRoadsters.com

the electric conversion components and parts. There are a number of these components, and they all have to work together as a system. As a result, you can't necessarily select one without considering the others and how they will (or won't) work in concert. The next seven chapters of this book will discuss, in detail, the major conversion components and parts, but, before we dive into any of that, it's useful to simply list the major components here and briefly describe their functions and key interactions. As it turns out, there are essentially seven major component groups, or subsystems, that you will need to select:

Motor and Adapter—There are a variety of factors that go into the selection of the electric motor you intend to use, including the type (AC or DC), the size and power rating of the motor, and the method of mounting the motor to the vehicle's transmission (i.e., the adapter). Also important is the overall system voltage and the type of controller you intend to use.

Battery Access

When looking at different vehicle body styles, consider how accessible the batteries and electronics will be for maintenance. A mini-truck conversion with the batteries in the bed will mean it's easy to service. As a result, you're more likely to check fluid levels in batteries on a regular basis. In contrast, a Porsche Speedster, with the batteries tucked into hard-to-reach nooks and crannies, will make maintenance difficult—and therefore less likely to occur.

Not all electric vehicles have an even number of wheels. This peppy little three-wheeler features a fiberglass body, removable hardtop, and side-by-side seating for the driver and passenger. Powered by an 8" DC motor and 72 volts of AGM-type lead-acid batteries, the car has a range of over 50 miles and can top out at 65 mph. With an onboard charger, the three-wheeler makes an excellent small commuter vehicle that can be driven for pennies per mile. Also, in many states, three-wheeled vehicles like this can be licensed as motorcycles, which can result in greatly reduced yearly registration and insurance fees. Courtesy Bill Provence and pureelectricvehicles.com.

Motor Controller—The motor controller is the electronic device that regulates the speed of the motor. The selection of the controller is dependent on the type of motor chosen, the system voltage, and intended use of the vehicle.

Batteries—There are a number of different types of batteries available from which to choose, including the ubiquitous flooded lead-acid and high-tech lithium-based units. Selecting a battery type is dependent upon requirements of vehicle range, speed, load-carrying capacity, and overall required system voltage.

Battery Charger—The selection of a battery charger is dependent upon the type of batteries chosen, the voltage and current availability of the supply electricity system, and the amount of time required to recharge the batteries every day.

Wiring and Switches—Wiring and switches need to be selected based primarily on controller type and system voltage.

Accessories—The selection of components such as DC-to-DC converters, heaters, air-conditioners, and power accessories are a function of system voltage, vehicle type, climate, and your own personal preferences.

Instrumentation—This area covers items such as voltmeters and ammeters, digital dashboard displays, and state of charge meters. The choice of instrumentation will be dependent upon vehicle type, system voltage, motor and controller type, and battery selection.

You may notice that the selection of the majority of these individual components depends strongly on the overall system voltage, which typically ranges from a low of around 48 volts on motorcycle and very small and lightweight vehicles, up to 300+ volts on high-performance sports cars and heavy vehicles. Most "normal" commuter EVs are operated at roughly 96 to 144 volts. Selecting a nominal system voltage is something that should be performed early in the design process. Generally speaking, higher voltage means higher top speeds; lower voltage systems reduce the acceleration and top speed of a vehicle, but also mean lower cost. It is strongly recommended that you look at what other builders of similar cars to the one you're planning have chosen, including an examination of what top speed and zero-to-sixty times they've achieved with the setup.

AC or DC?—Another major consideration at this step is the choice between an alternating current (AC)-based system and a direct current (DC) system, as this will drive a number of secondary considerations. Both types of systems have their merits, and the truth is that choosing AC or DC is a lot like choosing between a V8 and a V6 when shopping for a normal ICE-powered vehicle. An AC system will likely provide better overall performance and will allow relatively easy regenerative braking. The downside of an AC system, however, is cost of the motor and controller. The AC system will also likely require a higher overall system voltage, which means more batteries, which in turn means more money expended up front. The venerable DC system, like the V6, may not provide neck-snapping performance, but will be less expensive to purchase and there will be more

While it is certainly possible to convert a large vehicle such as this Chevrolet Suburban, it is not really recommended, especially for the first-timer. The sheer size of the battery bank, controller, and motor required to propel a 2-ton or greater vehicle like this is impractical for most people to undertake. A good rule of thumb when thinking about first-time conversion projects is this: Lighter is better.

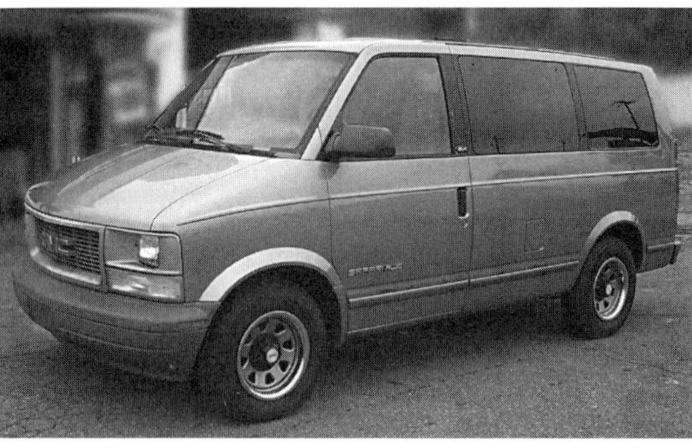

Another vehicle not recommended for the first time EV enthusiast is a large van, like the GMC shown here. In addition to requiring very large (read: expensive) battery banks, motors, and controllers, most vans come equipped with automatic transmissions. Automatics are generally much more difficult to adapt to electric power than a manual transmission. Making the van shown here even more challenging to convert is the fact that it's an all-wheel drive (AWD) model, which means extra weight, driveline losses, and inefficiencies. The lesson is this: Bigger cars don't necessarily mean longer range. In fact, the larger the car, the more batteries are required to operate it, which means more weight that is carried. Small, light, and aerodynamic are the best traits to look for in a first-time conversion vehicle.

people to draw upon for their help if/when things aren't working properly. For this reason alone, most beginners start with a DC system for their first conversion project.

As mentioned earlier, we will go into a lot more detail on individual component selection in the next seven chapters. That said, it is recommended at this point in the process that you begin researching EV parts suppliers. Batteries can usually be purchased locally, especially if you decide to use flooded lead-acid batteries. The rest of the EV components, however, will probably need to be ordered from online sources such as Canadian Electric Vehicles, the Electric Car Company of Utah, Electro-Automotive, EV Source, EV America, or KTA Services. Do your homework and talk to other customers who have used the supplier you're considering. Shop around and ask questions.

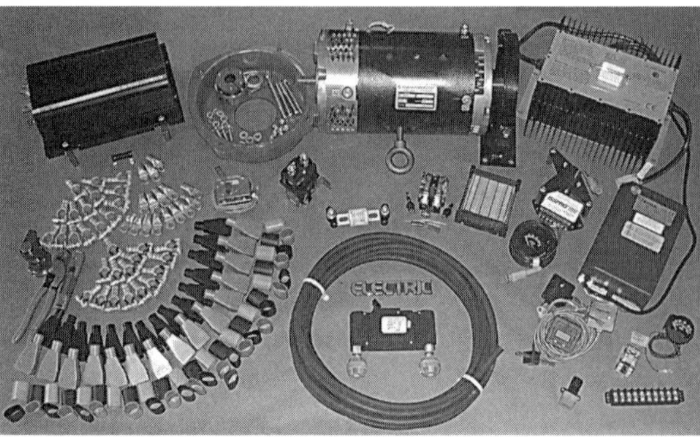

Electric vehicle components can be purchased either individually or in a kit, as shown here. The former method gives the builder the widest control over the project, while the latter simplifies the component selection process. Both methods can result in a first-rate conversion vehicle. Courtesy KTA Services.

Look for deals, but don't necessarily be swayed entirely by price; you often get what you pay for in the world of EVs.

5. Locate Work Space and Tools

Once you've decided on a vehicle and components, the next step is to secure a place to work on the car. The typical conversion project will take two to three months from start to finish if you're motivated and handy with tools but can take much longer if you're a procrastinator and/or unsure of which end of a wrench is which. Because of the time commitment, it's suggested that you find indoor space to work. A garage that can be closed and locked is the best solution; EV parts are expensive and, therefore, tempting targets for thieves, especially when left out for weeks or months at a time. If crime isn't an issue, temporary structures, such as tents or large sheds, make a good compromise. Even a covered carport is better than working in the open air; EV parts exposed to inclement weather prior to installation is a recipe for problems.

The work space ideally will have a concrete floor and will be equipped with reliable power outlets and bright lighting. There is nothing worse than dropping a critical bolt or connector onto a dirt floor in a poorly lit work space and not being able to find it. Even if it means cajoling your way into a friend's garage or perhaps renting temporary space, having a clean, well-lit, safe, and secure work space will make your entire conversion process much more enjoyable.

You also need access to a set of quality tools. The exact number and type of tools depends on the vehicle choice, the components selected, how much

Suspension Upgrades

When considering a vehicle for conversion, look into what aftermarket suspension and brake upgrades are available. The loads imparted on an EV from heavy lead-acid batteries can be substantial. Having the ability to install heavier duty springs and shocks, or larger, stronger brakes, should be a factor in your decision.

of the work will require fabrication, how much you intend to farm out to professionals, and whether or not an EV kit is employed. Some enthusiasts have gotten away with just basic hand tools and some standard power tools (e.g., hand drill, grinder), but others have found they needed things like a welder, drill press, or other large equipment. The trick is to mentally plan out as much of the work ahead of time in your head and make a list of the equipment you think you will need. The next trick is to then add these tools into your budget, even if the amount is just for a daily rental of a welder. Like the choice of work space, having the right tools on-hand will make your conversion project much more fun.

6. Convert the Vehicle and Drive

Okay, now you're ready to actually convert a car or truck to electric power. You've defined your goals and requirements. You've also decided on a budget, selected the best vehicle to convert, cleared out your garage and organized your tools. You also have a rough idea of all the components, parts, and subsystems that you need to source for the conversion, and you've begun talking to potential suppliers. In the next seven chapters, we will examine these individual parts and components in detail so that you can make informed choices and trade-offs prior to purchasing.

Want to build something with a little bit more zip than the average street-driven battery electric vehicle? How about an all-wheel drive race car that accelerates from rest to 60 mph in 4 seconds and can top out at over 120 mph? Palatov Motorsport in Oregon has built just such a beast, dubbed the DP1/e. The car is shown here in its natural environment, chasing down a formula race car. Courtesy Dennis Palatov and DPCars.net.

A look under the bodywork of the DP1/e race car. The AWD vehicle utilizes a 60 kW motor and custom controller. It may look radically different than the average EV, but all the basic components are present and accounted for, including motor, controller, and battery bank. Courtesy Dennis Palatov and DPCars.net.

Electric Motorcycles

Motorcycles make excellent electric conversion projects. They're lightweight and easy to work on, and can be built in a small shed or garage quite easily. They're also relatively inexpensive to convert; the typical motorcycle conversion can be performed for under $3000, including the cost of the donor motorcycle.

So what's different about a motorcycle conversion when compared to a car or truck? Not a lot, actually. All EVs use some kind of electric motor, a motor controller, and batteries. Cars and trucks require relatively large motors and controllers, and need relatively high battery system voltages. Motorcycles, because they're so lightweight, can get by with small motors and controllers. Instead of the typical series-wound DC motor, many motorcycle conversions utilize either brush- or brushless permanent-magnet (though others use series-wound DC motors). To control speed, the typical motorcycle utilizes a twist-grip throttle (pot box) assembly to vary the motor controller's commanded speed.

Because a reverse gear isn't required and high performance can be achieved with a single drive ratio, a major difference between a typical car conversion and a motorcycle conversion is the latter doesn't usually use a transmission or gearbox. Most motorcycle conversions simply rely on a direct chain- or belt-drive connection between the motor and the rear wheel.

Motorcycles can also run on much lower system voltages; the typical conversion is built around 48-60 volt battery banks. Because of this, many builders find that they can afford high-performance lithium-based battery systems, which result in very good range and high performance.

Like a car or truck, the typical street-driven motorcycle conversion requires a DC-to-DC converter, which is used to power the standard lights, turn signals, horn, and so on. Similarly, some motorcycles use onboard chargers that allow opportunity charging, while others keep the charger at home, like many car and truck conversions.

Motorcycles make excellent first-time conversion projects, as they are lightweight, easy to work on, and can operate on relatively low-voltage power systems. Roller (i.e., non-running) motorbikes can often be found for very little money on craigslist.org and other online sources. Older Honda Interceptors, like the one shown here, are popular among EV enthusiasts. Also popular are Honda Shadows, Kawasaki Ninjas, LIFAN motorcycles, Suzukis GS-bikes, and Yamaha FZRs and R1s. Two-stroke dirt-type bikes, such as Honda CRs, Kawasaki KXs, KTMs, Suzuki DRs, and Yamaha YZs, have all been converted into strong-running BEVs. Some enterprising builders have even converted Harley Davidsons to electric, though you might first check with your local biker gang to see if they approve.

An off-road dirt bike motorcycle that has been converted to electric power. Courtesy Bob Simpson.

EV CONVERSION COMPONENTS

Chapter 4
Electric Motors

"Before I put a sketch on paper, the whole idea is worked out mentally...My first electric motor [has been] developed in exactly this way."
—Nikola Tesla

An electric motor is the heart of an EV. One of the most popular series-wound DC motors used in EV conversion projects is the ubiquitous Advanced DC FB1-4001A 9.1" motor. Weighing in at roughly 150 lb, this proven motor can produce 28.5 continuous horsepower at 144 volts input. Courtesy KTA Services.

The heart of any electric vehicle project is its electric motor. Choosing the right motor is key to the success of an EV conversion. If the selected motor is too small, the car will be slow, will underperform, and may even be unsafe for use on busy streets. On the other hand, if the motor is sized too large for the application, the system will be heavy, inefficient, and costly.

This chapter is intended to help you wade through the different options and choices when it comes to selecting an electric motor for your conversion project. We will start the learning process by reviewing basic electromagnetism, which is the physical phenomenon that allows a motor to operate. We will then examine the construction and operation of a simple DC motor. Following this, we will then look at the major types of electric motors that are suitable for use in an EV, including a discussion of their relative pros and cons. Finally, we will look at the options and methods used to physically adapt and mount an electric motor to your vehicle.

Electromagnetism Basics

Before we can understand how an electric motor works, we need to first grasp how an electromagnet operates. This in turn begins with an examination of permanent magnets, like the type you might have on your refrigerator door that is used to hold a note or piece of paper in place. Magnets like this will attract and hold to metal objects that are near or in contact with the magnet. This attraction is possible because the space around the magnet contains something called a magnetic field, which is created by the movement of negatively charged electrons in and around the magnet. Electrons have mass and a negative charge and, in magnetic metals, are unpaired and spin, which create magnetic lines of force that can attract or repel.

The magnetic field of a permanent magnet is demarked by the boundaries of its so-called magnetic lines of force. These lines of force are the north-to-south lines in a magnetic field that we learned about in high school science class. Magnetic lines of force, or *flux lines*, start at the north end of a magnet and flow to the south end, just like they do on the Earth's axis. Any magnet will try to set itself parallel to the lines of flux. This is how magnetic compasses work; they align themselves with the earth's magnetic field. This is also why magnets repel or attract other magnetic materials, depending on the north-south position of the magnetic flux lines.

Magnetic flux lines can flow from the north pole of one magnet to the south pole of another. These flux lines also characteristically try to be as short as possible in their continuous path from a north pole to a south pole. Therefore, the flux lines that pass directly from the north end of one magnet to the south end of the other will tend to draw the two magnets together. In a similar manner, when two like poles of two magnets are brought together, the flux lines form a kind of buffer between them, which acts to separate the two magnets. When it comes to magnets, opposite poles attract, and like poles repel each other.

We can create temporary "electromagnets" that behave just like permanent magnets by way of electrical current, which is essentially the movement of negatively charged electrons. If

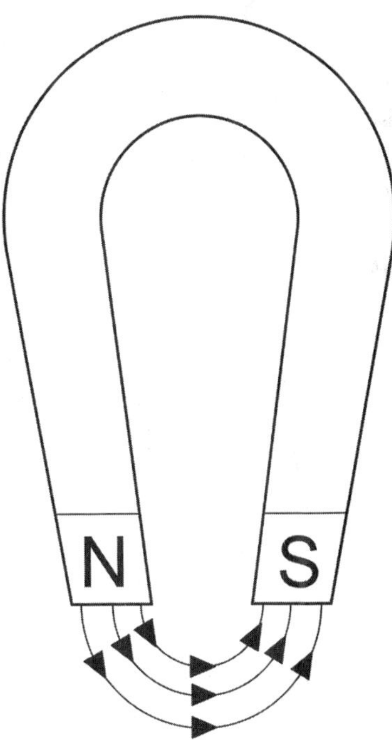

The lines of flux of a permanent magnet are always from its north pole to its south pole. In this horseshoe magnet example, the two magnet ends are nearby each other and the flux lines have a relatively direct route to travel.

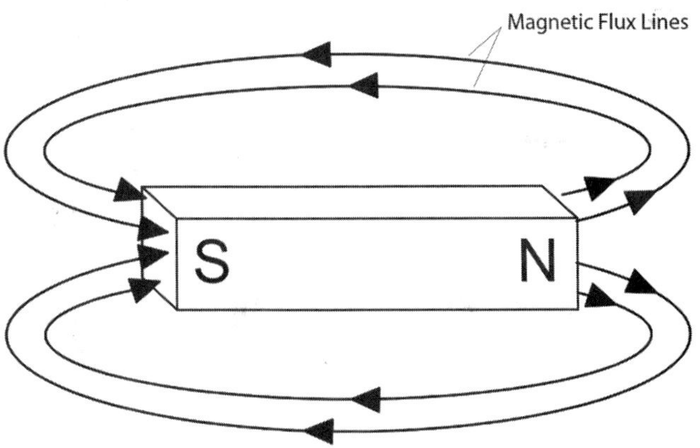

The magnetic lines of flux of a straight bar magnet travel outward from the north end and loop back to the south end, where they reenter the magnet.

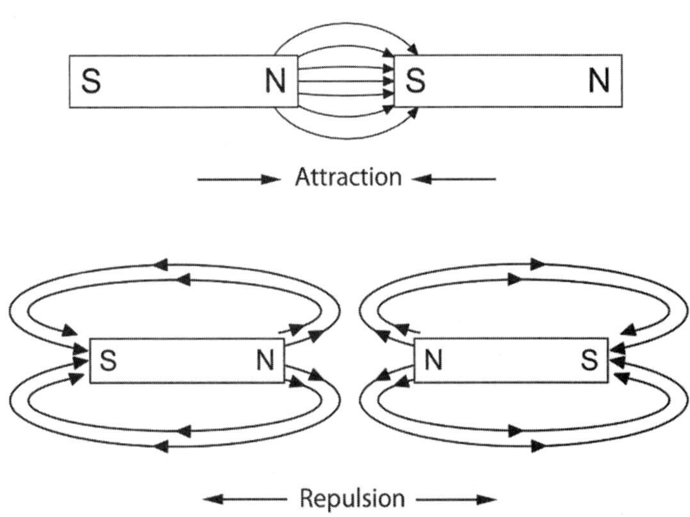

Both permanent and electromagnets behave the same when they are brought into close proximity with another magnet (either permanent or electromagnet). The north end of a magnet will attract the south end of another magnet. The north end will also repel another north end of a different magnet. Opposite ends attract, and similar ends repel.

we pass an electric current through a straight piece of copper wire, for example, a magnetic field is generated that is proportional to the amount of current. This field can be thought of as concentric lines of force (flux lines) that encircle the wire. The stronger the electric current, the stronger the resulting magnetic field. If we reverse the direction of the current, the field shape remains the same, except that the lines of force reverse direction. If we switch off the current, the magnetic field disappears; when we switch the current back on, the wire becomes magnetized again.

The magnetic lines of force in an electromagnet act in exactly the same manner as those found surrounding a permanent magnet; i.e., they attract and hold to other magnets and to metal objects. If we want to amplify the wire's magnetic field, we can increase the current flow through the wire. We can also achieve a similar effect by simply coiling the wire into loops. Doing this causes the magnetic flux line to add in strength without having to increase the current. The more loops, the stronger the resulting magnetic field.

Another similarity between permanent and electromagnets is how they behave when they are brought into close proximity with another magnet (either permanent or electromagnet). Just like a permanent magnet, the north end of an electromagnet will attract the south end of another magnet. The north end will also repel another north end of a different magnet.

This behavior of electromagnets, such as opposites attracting and field strength increasing with loops, or windings, is what allows electric motors to work so efficiently. Let's take a look at a simple DC motor to see how this all works together.

How an Electric Motor Works

There is a wide variety of types and configurations of electric motors, but perhaps the simplest of all is a brush-type DC motor. The following description shows how this type of basic motor is constructed and operates.

As we saw in the previous section, we can create an electromagnet by coiling wire in a series of loops. The more loops, the stronger the resulting magnetic

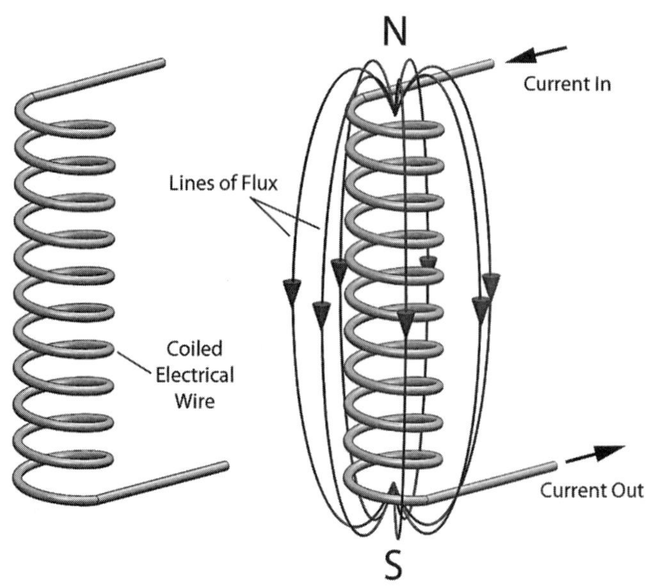

Wire wound into a coil can behave just like a permanent bar magnet if an electric current is passed through the windings, including north and south poles and magnetic flux lines. The result is something called an *electromagnet*. Stronger current and/or more loops per inch can increase the strength of the magnetic field.

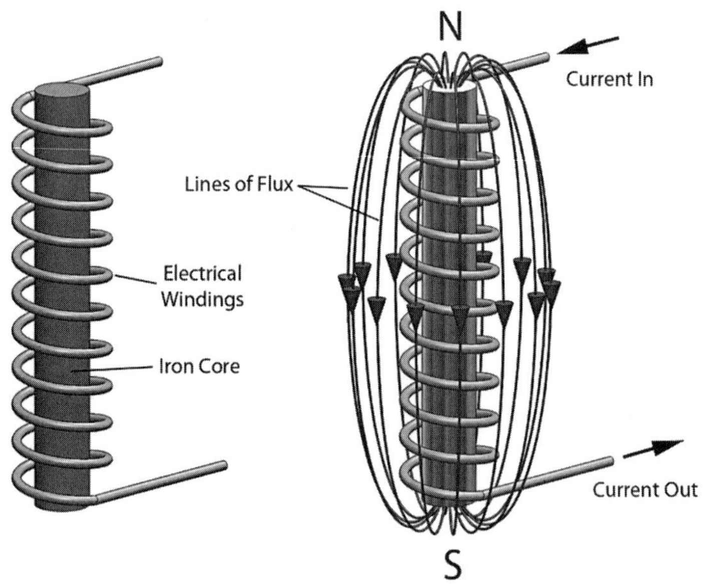

The strength of an electromagnet can be increased with the addition of an iron core placed inside of the windings, which essentially is how most modern electric motors are built.

> Electric motors are normally rated at their point of maximum efficiency. A motor may be capable of two to five times its continuous rating, but usually only for a few minutes, such as when passing or climbing a hill. In comparison, internal combustion engines are rated at their peak horsepower output. This is the reason it's difficult to compare horsepower ratings of EVs to ICE-powered vehicles.

field. We can further strengthen this field by coiling the wire around a core of iron. If we were to build this type of electromagnet, we would have a motor armature. If we then took this armature and placed a pivot or bearing at the center of the electromagnet that allowed the entire unit to rotate, we will have created something called an electric motor rotor.

Let's imagine that this simple rotor is laying flat on the ground and is free to rotate either clockwise or counterclockwise about the central pivot point, much like the spinner used in a child's board game. We can turn the rotor by hand about its pivot so that if we were to energize the coil with an electrical current, the north pole would be pointed up in the 12:00 direction and the south pole pointed down in the 6:00 direction.

With the electromagnet in this position, but switched off, we can then position and fix two permanent magnets near the ends of the electromagnet. One of these permanent magnets is placed such that its north pole is nearest the end of the electromagnet that is at 12:00. The other permanent magnet is placed at the other side of the rotor with its south pole nearby the electromagnet at 6:00.

Now, if we were to energize the electromagnet by attaching a battery to the coil and flowing electricity through it, we would suddenly have a situation where a north electromagnetic pole was in close proximity to the north permanent magnet pole, and a south electromagnetic pole was similarly in close proximity to the south permanent magnet pole. Because like poles repel each other and the rotor is free to rotate on its pivot, the electromagnet would move. In fact, because like poles repel and opposites attract, the electromagnet would spin 180 degrees and then stop, with its north pole now pointed down at 6:00 (drawn to the south pole of the permanent magnet there) and its south pole pointed up at 12:00 (drawn to the north pole of the permanent magnet located there).

Commutation—This is all fine and dandy, but we don't want the rotor to stop spinning after just 180 degrees of rotation. We want it to continue rotating around and around in continuous motion. The trick to doing this is to suddenly reverse the current of electricity flowing through the electromagnet just as it completes 180 degrees of rotation. This act of reversing the current is called "commutation," and the result is that what was the north pole of the electromagnet suddenly becomes energized as a south pole. Similarly, the south pole quickly switches to a north pole, and, because the permanent magnets haven't moved, our rotor suddenly is in a situation again where like pole are next to each other. The result of course is that the rotor spins another 180 degrees back to its original location. If we switch the flow of electricity again and again in this manner every 180 degrees of rotation, the electromagnet rotor will continue to spin around and around, producing continuous motion that we could harness to do real work.

Now, astute readers will question how it's possible to connect a battery and wires to a continually spinning rotor. Also, how do we commutate (i.e., switch) the flow of current in a manner that not only does so quickly, but does so at the precise correct time during rotation to keep the rotor moving smoothly and continuously? The solution is by way of something called motor brushes.

Motor Brushes—Motor brushes are electrical contactors that rub (i.e., brush) against the shaft of the rotor. Electricity can flow from a fixed, stationary source (such as a battery), through a brush, and onto the shaft of the rotor. By employing two brushes on opposite sides of a shaft we can flow electricity onto and off of the spinning rotor. Also, by separating the two coil leads 180 degrees apart on the rotor shaft we are able to create a self-commutating system. Electricity flows through one brush onto the rotor and flows in one direction through the electromagnet coil and then out and across the second brush. When the rotor has turned 180 degrees, electricity, which is continuing to flow onto the rotor by the first brush and off the rotor by the second brush, now passes in the opposite direction through the coil, thereby switching the north and south poles. When the rotor spins an additional 180 degrees, the flow of electricity is reversed again. Voila, we've created a very simple brush-type DC motor.

More advanced versions of this type of motor incorporate multiple magnet poles and other tricks to smooth out the movement of the rotor and enhance its power output. Further, there are other, more advanced permutations and variations of

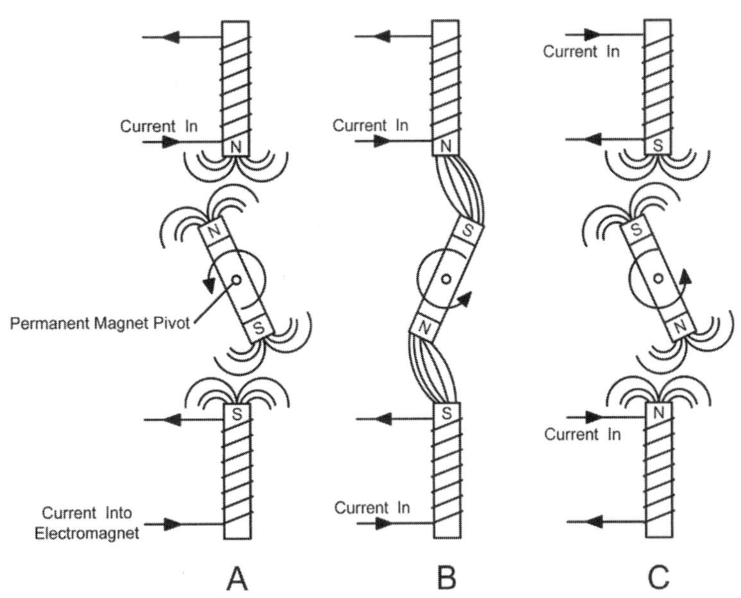

A simple motor can be constructed by putting a pivot axle in the center of a permanent bar magnet (rotor). If we put electromagnets (stators) on either side of the rotor and energize them, the north end of the rotor will be repelled from the north end of one of the electromagnets, and the south end will also be repelled from the south end of the other electromagnet. As the rotor turns about its pivot axle, the north end will move closer to the south end of the lower electromagnet and will be attracted to it. The same will happen to the south end of the permanent magnet, and it will be drawn to the north end of the upper electromagnet. As the opposite ends get closer to each other, we can reverse the current direction of the electromagnets, which results in repulsion at both ends. By switching the current direction over and over, we can cause the inner permanent magnet rotor to continue turning about its pivot axle.

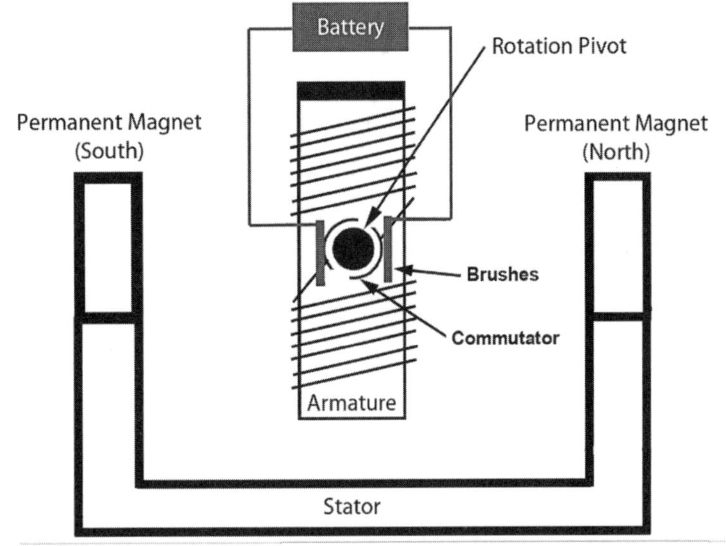

Instead of a rotating permanent magnet and fixed electromagnet stators, many simple DC motors utilize stationary permanent magnets and a rotating electromagnet. Brushes that rub against commutator bars on the rotor cause the direction of the electrical current to switch every 180 degrees.

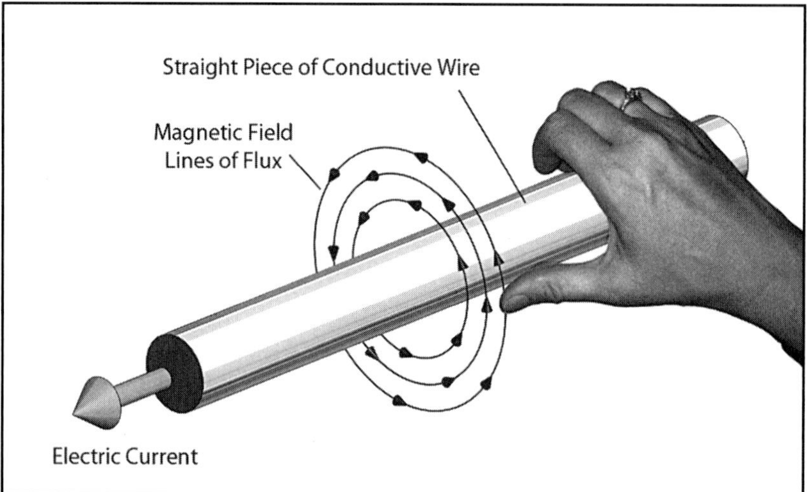

When a current passes through a conductor, lines of magnetic force (flux) are generated around the conductor. The direction of the flux is dependent upon the direction of the current flow. If you think of current in the conventional sense (i.e., positive to negative), the Right-Hand Rule can be used to determine which direction the magnetic field lines are pointed. Simply take your right hand and grip the wire with your thumb pointed in the direction that the current is flowing. Your other four fingers will point in the direction of the flux lines.

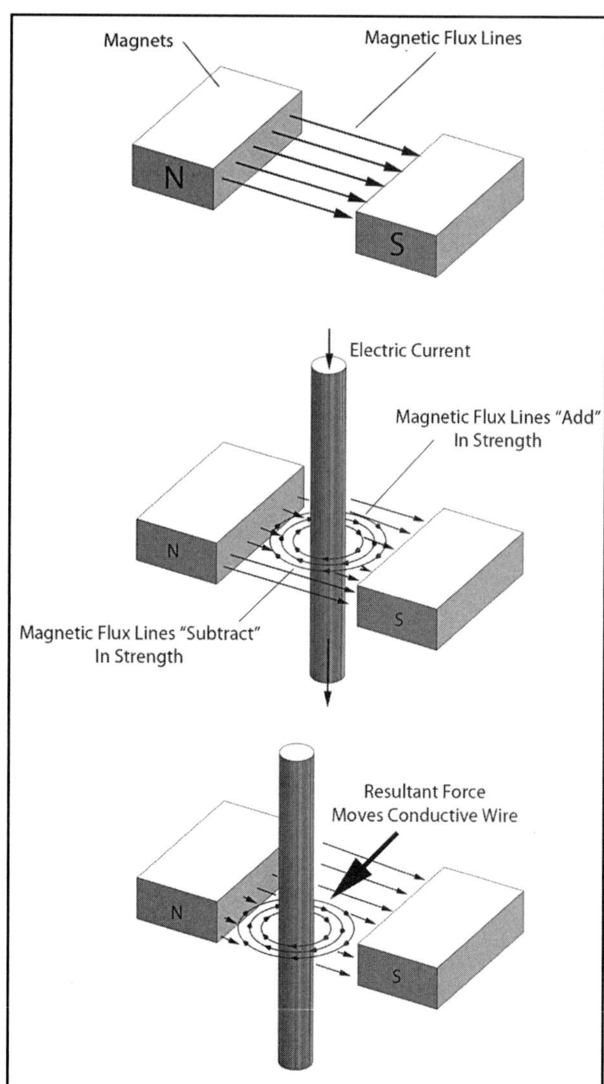

If we place a current carrying conductor in the air gap of two magnets, the lines of flux will be affected. On the side of the conductor where the flux lines oppose each other, the magnetic field will be made weaker. On the other side, where the lines of flux are in the same direction, the magnetic field will be made stronger. Because of the resulting imbalance of field strength from one side of the conductor to the other, the conductor will be pushed into the weaker field.

Power to Weight

A rule of thumb says 10 continuous horsepower is required for each 1000 lb of vehicle weight (after conversion to electric power). A 2000-lb EV would therefore require a motor rated at 20 horsepower. Even more would be required if the vehicle were required to be operated on hilly terrain or at unusually high speeds.

electric motors, including brushless, series-wound, and even AC motors. We will look at these different types and options in the next section.

Motor Types, Sizes, and Ratings

Electric motors are ubiquitous. They are found in applications ranging from tiny watch motors to giant units used to power ships and locomotives, with nearly every conceivable size and shape in between. They are also constructed in a number of different ways and operate on a variety of different principles. Motors can be brushed or brushless, powered by direct current or alternating current, and can operate at fixed speeds, variable speeds, or can simply be used to move from one position to another in discrete angular steps. Servo motors can be used to position equipment, but then so-called stepper motors can also serve a similar function, yet they operate on a different principle. There are even linear motors and partial-segment designs available for specialty applications.

Sorting through all these different types of motors can be confusing to the first-time EV constructor. That's the bad news; the good news is there are only a few types of electric motors that are important to consider for most EV conversion projects. In this section, we will look at the major categories of EV-capable motors and discuss their relative pros and cons. Let's start the discussion by classifying electric motors into two separate major types: DC and AC.

HP Math

Horsepower requirements of a vehicle increase as the cube of speed. Double the speed and you need eight times the horsepower to overcome aerodynamic drag.

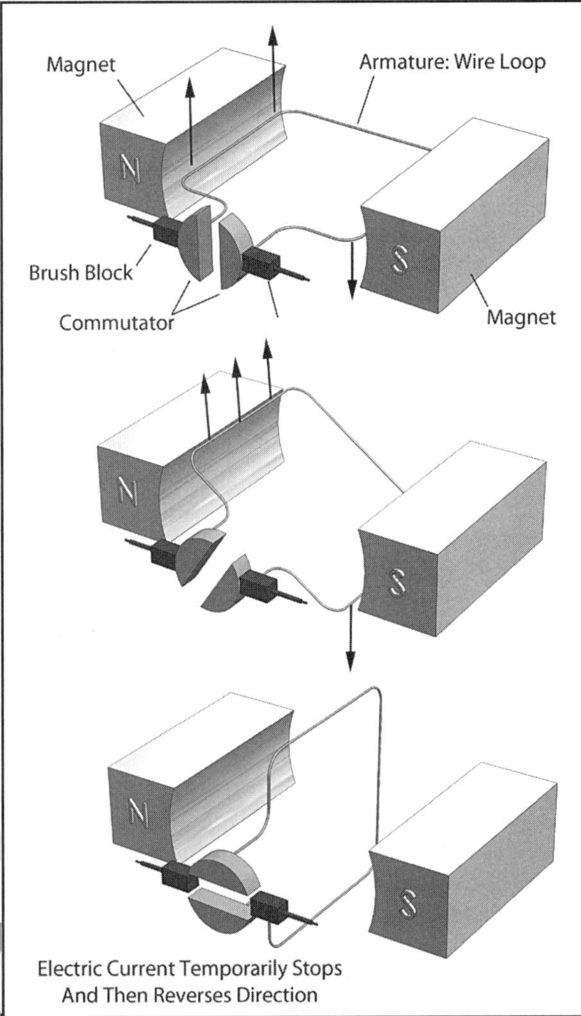

We can create a simple DC motor by creating a loop of wire capable of carrying a current and rotating. We can also place this assembly inside of two permanent magnets (one north and one south). A DC source, commutators, and brushes complete the circuit. When current passes through the wire loop, the lines of flux above and below each side of the wire will be affected. The result is a force acting upward on the wire near the North permanent magnet and downward on the wire near the South magnet. When the loop has rotated 90 degrees, the commutator temporarily stops supplying electrical current, but the loop continues to move due to its own inertia and momentum. The commutator then contacts the brushes again, but this time the current is in the opposite direction, thereby continuing to pull the wire loop around on its rotation axis.

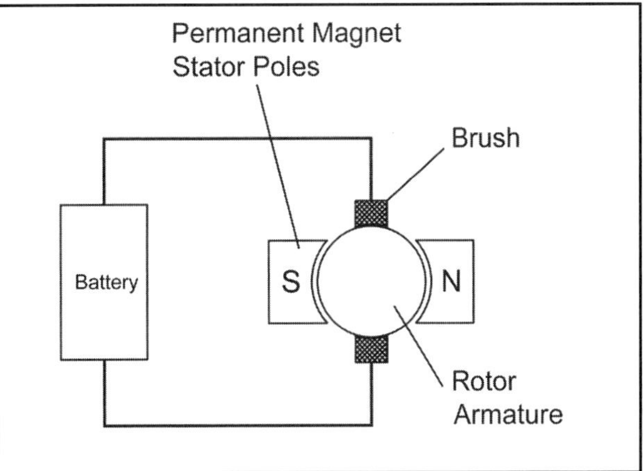

A simple permanent magnet motor circuit. Electricity flows from the battery to the rotating armature via stationary brushes that run up against commutator contacts on the armature. The permanent magnets create a stationary field against which the rotor electromagnet operates.

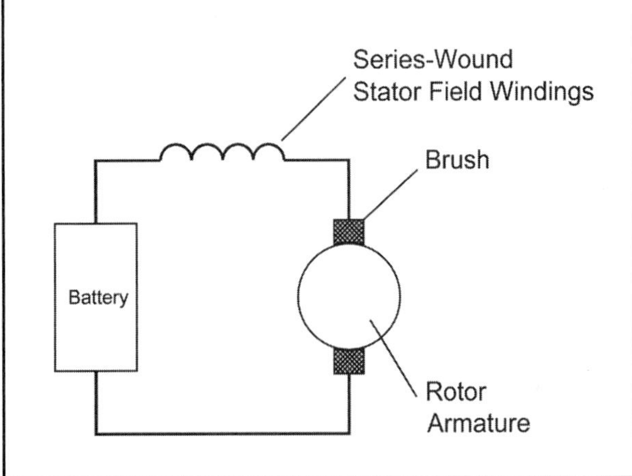

In a series-wound circuit, the stator field windings are supplied electricity in series with the rotor electromagnets. Contact brushes periodically reverse the rotor electromagnet just as they do in a permanent magnet design.

DC Motors

Permanent Magnet—The simple example described in the previous section was a DC-type motor. Specifically, it was a permanent magnet, brush-type DC motor that used a single electromagnet (with one north and one south pole) armature for the rotor and two separate magnets on the stator side. This motor is called a DC motor because the power supply connected to and that energizes its electromagnet is a direct current source (e.g., a battery). It is called a permanent magnet DC (PMDC) motor because it utilizes permanent magnets on the stator side of the motor. And it is called a brush-type because it employs brushes that contact a split-ring commutator to switch the magnetic field of the armature every 180 degrees of rotation.

This motor is a very simple device; in fact, it's too simple to be practical for use in an EV. The limited number of poles and magnets means the motor would have a very strong cogging action and would not operate in a very smooth manner. It would also be relatively weak and not produce very much useful torque. The majority of DC motors have

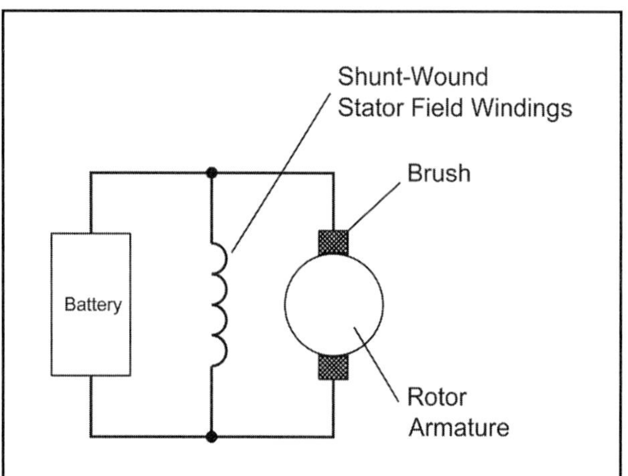

In a shunt-wound circuit, the stator field windings are wired in parallel with the rotor electromagnets.

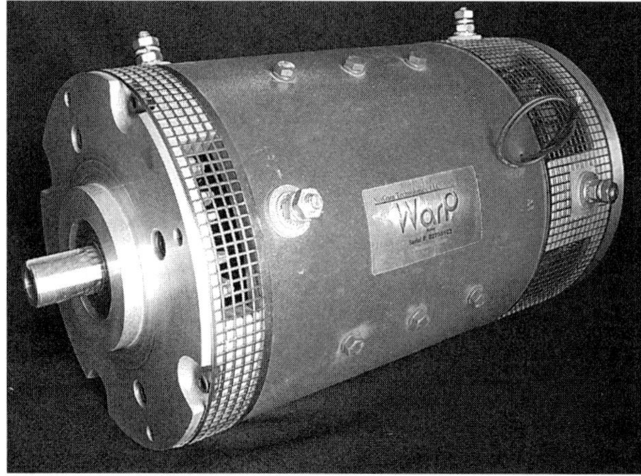

Netgain Technologies Warp 9 is a popular 9.25" diameter, series-wound DC motor with a double ended shaft. Netgain makes a range of Warp motors, from a modest 8" unit up to a monstrous 13" motor. Courtesy Canadian Electric Vehicles.

Permanent magnet motors make excellent drive units for small EVs, such as motorcycles and scooters. Shown here is an 8" diameter Mars permanent magnet motor. Typical operating voltage for this motor is 48–72 volts.

more than one electromagnet armature and more than just two stator magnets, thereby smoothing out the power pulses and allowing higher torque outputs.

For very lightweight EV applications, such as motorcycles and bicycle conversions, a permanent magnet motor is often a very good solution. PMDC motors are small and lightweight, which makes packaging them in small spaces relatively easy. Also, the torque requirements for these smaller EV applications are not usually large, which means that a moderate sized PMDC can usually be found for relatively low cost. PMDCs are also slightly more efficient than other types of DC motors, which helps when you're trying to squeeze the last dollop of performance out of a lightweight bike or motorcycle conversion. The downside of these types of motors are they aren't available in large horsepower sizes, which means they aren't practical for car and truck conversions. They also produce a significant amount of back EMF, or braking action, as the rotation speed increases above a certain upper threshold level. Control of the speed of a permanent magnet motor is usually carried out by way of voltage control; the higher the voltage, the faster the motor rotates—up to the point where back EMF grows too large and the motor starts acting as a generator, thereby slowing the motor.

Field-Energized DC—The other major type of DC motor is the field-energized type. Like a PMDC, a field-energized motor utilizes electromagnets in the rotating armature. Unlike a PMDC, however, a field-energized motor does not utilize permanent magnets in the stator; instead, electromagnets are employed. The majority of car and truck EV conversions today are built using field-energized DC motors. This is primarily because this type of motor can produce more horsepower and torque than a PMDC motor. Field-energized motors used in EVs usually fall into one of three major types: series-wound, shunt-wound, and compound-types.

By far, the most common type of motor used in car and truck conversions is the *series-wound* motor. The term *series* describes the field and rotor windings, which, as the name suggests, are wired in series. In other words, current from the motor controller flows first through the stator field windings, and then through the rotor windings. The chief advantage of series-wound motors is their high torque abilities and their overall simplicity. Series motor controllers are available from many different suppliers, and are available with a variety of options. They can also be mixed and matched relatively easily with different series-wound motors.

(We will look at controllers in more depth in the next chapter.) Because of the popularity of series-wound motors, there is a corresponding wealth of technical and hard-won practical knowledge available among EV enthusiasts. This means that when you have a problem installing or troubleshooting a series-wound motor, there is a large pool of technical knowledge available to help you work your way through the issues.

The chief disadvantage of series-wound DC motors is that they cannot (easily) provide regenerative braking. Some enterprising enthusiasts have built regen systems, but this is not generally recommended for the average EV constructor. Series-wound motors also have a unique problem of being easily over-revved if the controller is not designed to detect and limit over-speed.

Unlike AC motors, which are typically specified in terms of their power ratings, series-wound DC motors are usually classified (and marketed) by their nominal housing diameters. While this may seem odd at first, most EV parts suppliers will be very familiar with series-wound motors and sizes and will easily be able to recommend a motor size (and controller) for your application. Typically, 6–7" diameter motors are best suited to very lightweight vehicles, motorcycles, and golf carts. For "average" car and truck conversion projects, 8" or 9" motors will provide good performance. For larger vehicles or race applications, 11" or larger motors can be used. The two most popular modern lines of series-wound motors are the Advanced DC and the WarP motors. Used motors by Baldor, General Electric, Prestolite, and other brands can often be found for sale on the Internet. The downside of using these older motors is they may be inefficient, incompatible with modern electronic controllers, and/or difficult to mount and interface to your vehicle. Forklift motors are also sometimes used in conversions but are generally not recommended, both for the reasons listed previously for older style motors, and also because they are not always designed to run at continuous high speeds. Forklift motors are also usually much heavier than comparable purpose-built EV motors by Advanced or WarP.

In contrast to series-wound motors, *shunt-wound* DC motors are designed to have shunt, or parallel-wired power applied to the stator field and rotor windings. In other words, current from the motor controller is divided into two paths; one that passes through the stator field windings, and the other that is directed to and through the rotor windings. This type of motor, while seemingly offering some benefits over series-wound motors, is not frequently used in modern EV applications. The upside of shunt-wound motors is their ability to provide relatively easy regenerative braking. The downside, however, is the need for a more sophisticated motor controller (read: expensive and/or difficult to source). Also, shunt-wound motors are best suited to constant-speed applications, which generally do not fit the typical electric vehicle requirements.

The third common type of field-energized DC motors is the *compound-wound* motor. Like the name suggests, a compound-wound DC motor has both series- and shunt-wound field windings, which presumably offers the best of both worlds for the applications. Unfortunately, modern controllers that are suitable for vehicle usage are both expensive and difficult to source.

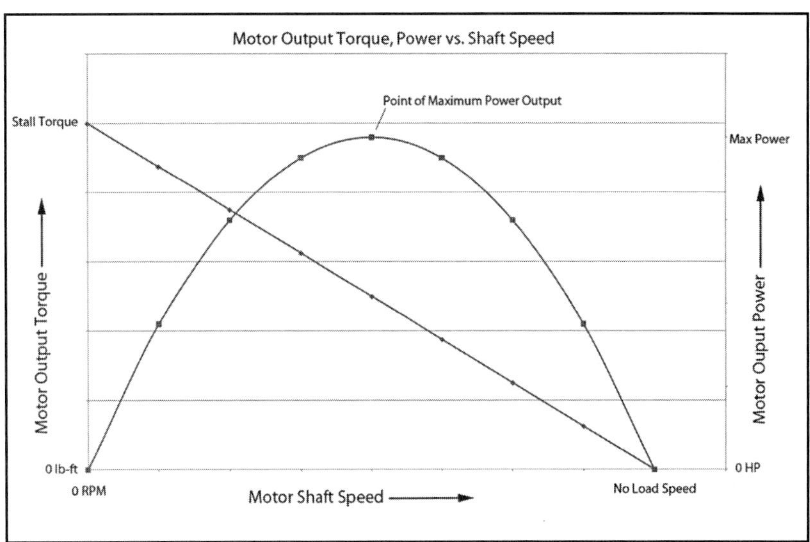

Many first-time riders in electric vehicles are surprised by how quickly the vehicles can accelerate from a standing start. Acceleration of an EV is directly proportional to the torque output of the motor, and electric motors typically make maximum torque at or near zero rpm. In other words, EVs accelerate most strongly when they're just starting off from a stop. Note that the point of maximum power occurs in the middle of the speed range of the motor. While acceleration is a function of torque, top speed is a function of power.

A brushless DC hub motor installed on the author's electric bike.

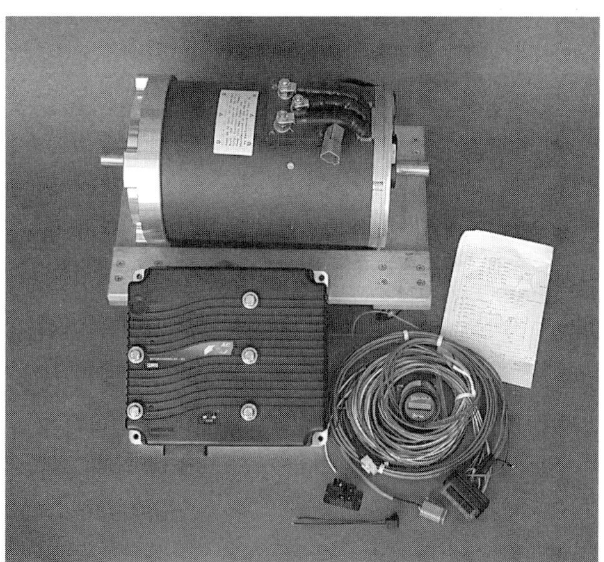

Shown here is a high-performance AC motor and controller kit. The HPGC AC50-01 motor is capable of producing 50 peak horsepower at 96 volts. The controller is a Curtis 1238-7501 that can provide regenerative braking. Courtesy KTA Services.

A Siemens brushless 3-phase AC induction motor that is water cooled. The unit is approximately 10" square and 27" long, and weighs 200 lb. Courtesy Bob Simpson.

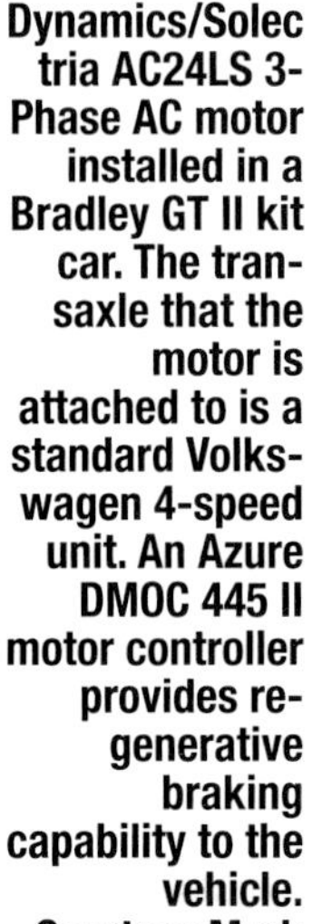

An Azure Dynamics/Solectria AC24LS 3-Phase AC motor installed in a Bradley GT II kit car. The transaxle that the motor is attached to is a standard Volkswagen 4-speed unit. An Azure DMOC 445 II motor controller provides regenerative braking capability to the vehicle. Courtesy Mark Bush.

The same Siemens AC motor in the process of being mated to a BMW transmission. Both an adapter plate and shaft adapter had to be custom designed and built for this particular application. Courtesy Bob Simpson.

AC Motors

If you hang around EV enthusiasts long enough, you will hear very passionate arguments why (or why not) AC motors are superior to DC motors in electric vehicle applications. Some of the arguments are valid, while others are simply opinion and/or old wives tales. Let's look at what comprises an AC motor and how it is rated, and list the benefits (and drawbacks) it has over a comparable DC motor.

Like its name suggests, an AC motor uses alternating current to energize the stator and rotor field windings and power the motor. (Note that many classic DC-type motors can be run on AC power; notably, so-called universal motors.) Unlike DC motors, AC motors are not generally classified by their physical size, but rather by their rated continuous power output. For average size cars and trucks, 15–20 kW is usually more than sufficient. For heavier applications, 30 kW motors are often required. Some high-performance vehicles use 50 kW or higher motors with excellent results.

One of the major advantages AC motors have is their ability to provide regenerative braking. This means that if you live in a hilly area and/or frequently drive in heavy stop-and-go traffic, you will be able to recoup much of the energy you normally waste as heat generated by the brakes. AC motors also have very favorable torque vs. rpm characteristics, which make them well suited to applications that require a mix of in-town and highway driving. For these and other reasons, large-scale manufacturers such as Ford, GM, and Toyota are developing AC systems for use in the next generation of commercial electric vehicles.

The downside of AC motors is their cost, or, more accurately, the cost of their support systems. AC motors by themselves are relatively inexpensive, but the need for expensive inverting controllers,

An adapter plate and flywheel bolted to an Advanced DC 8" motor. This particular adapter was built by KTA/Electro-Automotive and is designed to mate the motor to an MG Midget 4-speed transmission. Not yet installed on the flywheel in this photograph is the MG clutch assembly. Courtesy Mark Hayes.

Attaching an electric motor to an existing vehicle transmission can be carried out by way of homemade adapter plates and hub connectors. The simpler means, however, is via pre-engineered adapter kits. Ensuring precise and accurate alignment of the motor to the transmission, accounting for the flywheel and clutch requirements, and ensuring sufficient strength is provided means that designing and building your own adapter is not for the beginner. Vendors have already done all the homework necessary for most popular transmissions and transaxles. Why reinvent the wheel when you don't have to? From left to right in this image are commercially available adapter plates for Ford Duratec, Porsche, and Nissan transmissions, respectively. Courtesy Canadian Electric Vehicles.

coupled with the fact that most AC systems operate on higher nominal system voltages than comparable DC systems (and therefore require more batteries), often pushes the price of an AC system out of the reach of a budget-conscious EV constructor. The prices of AC systems are in fact coming down, but at the time of this printing they remain two to three times as expensive as a similar DC system.

Another point to consider when choosing between AC and DC-based systems is the fact that AC motors almost always have to be matched and sold with their corresponding controller. DC systems are much more flexible, allowing the usage of one brand of motor with another brand or type of controller. This is either a good thing or a bad thing, depending on your personal point of view.

The two most common AC motors used in electric car conversions are the Azure/Solectria type and quality European motors imported by manufacturers such as Metric Mind.

Motor Mounts and Adapters

Once you have selected an electric motor for your EV application, the next step is attaching it to the transmission or transaxle. To make this connection, there are usually three items required: a bellhousing adapter plate; a hub, or shaft coupling; and the motor mounts themselves:

Bellhousing Adapter Plate—The primary function of the bellhousing adapter plate is to securely attach the stationary stator side of the electric motor to the fixed, or stationary portion of the transmission (which is known as the bellhousing). The standard means of attaching the motor stator to the bellhousing is by way of a flat plate of aluminum or steel that has a large center hole through it for passage of the motor shaft. The plate must be accurately machined so that it precisely centers the electric motor onto the transmission. Getting this even a little bit wrong will cause excessively large loads on the electric motor bearings. The plate must also be strong and stiff enough to transfer the reaction torque loads that the motor housing imparts into the transmission during acceleration.

Hub Coupling—To connect the rotating output shaft of the electric motor to the input shaft of the transmission, a coupling is required. The design of this couple is dependent on a number of factors, including the type of transmission shaft (e.g., splined or keyed), and whether a clutch is going to be used or not. Couplings can be made from aluminum, but often steel is used because of its superior strength and durability.

Motor Mounts—Some builders forgo use of separate motor mounts, relying instead on just the bellhousing adapter plate interface to support the weight of the motor. While the motor may be designed to be supported in this manner, it's not a recommended solution, as there needs to be a means of fully counteracting the torque that the electric motor produces. Again, the bellhousing adapter plate provides some of this function, but this means that the transmission must be fully fixed to the chassis to transfer the forces and torques into the frame. Most transmissions and transaxle mounts are not designed for this purpose, so additional means need to be employed to beef up the attachment between the transmission and the frame. A better solution is to simply use a mounting bracket near the rear of the motor that can both carry some of its load, and also fully react

When shopping for a motor adapter kit make sure that it comes with everything you will need to bolt the motor to the transmission in your vehicle. This includes all the bolts, keys, and other necessary hardware. Courtesy KTA Services.

An 11" high-power WarP-11 motor attached to a BMW transmission, prior to installation in an EV. Courtesy Tim Catellier.

An adapter kit for bolting an 8" or 9" DC motor to a GM-type transmission. This adapter fits the transmission of the popular Chevrolet S-10 truck, as well as a host of other GM vehicles, including the Pontiac Fiero and Chevy Astro Van. Courtesy EVSource.com.

the torque of the motor. If possible, it's best to use a bracket that interfaces to the factory ICE mount locations in the engine bay. These hard points are designed for the loads and stresses imparted by an engine, so they are naturally strong enough to carry the loads and stresses from the electric motor replacement.

For some popular vehicles, such as Chevy S10s, Porsche 914s, Volkswagens, and the like, there are companies available that have complete adapter kits

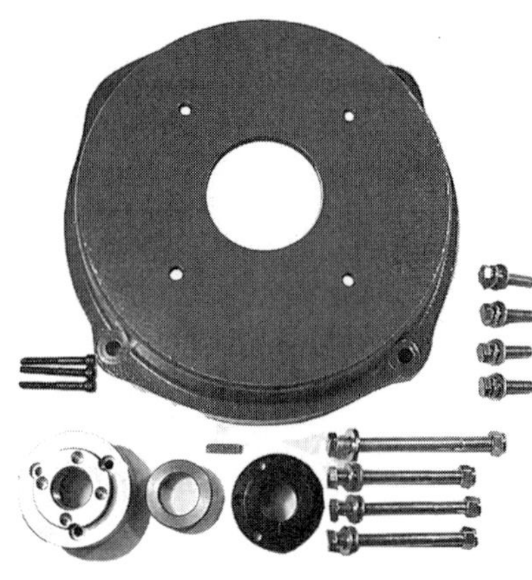

A Volkswagen adapter plate and coupling kit. Courtesy EVSource.com.

(bellhousing adapter, hub coupling, and motor mounts) available for purchase off the shelf. You simply order the appropriate adapter equipment for your motor and vehicle combination, and it will be delivered in a few weeks.

If you're converting a vehicle that doesn't have a kit available for purchase, you will need to have the mounts and adapters constructed. For many applications like this, it's possible to have a specialty firm, such as Electro Automotive, build an adapter to your needs. This often requires either carefully measured and detailed drawings supplied by you, or you need to produce a paper rubbing of the transmission bellhousing so that they can use it as a template for an adapter plate.

Transmissions, Flywheels, and Clutches

Electric motors generate full output torque at zero rpm, but most electric vehicles still require some type of gear reduction device to multiply this torque output. For cars and trucks, the most obvious means of achieving this gear reduction is via the original manual transmission. Note that we will be assuming the use of a manual transmission here in this section. Automatic transmissions can be used in electric vehicles, but they are not generally recommended. Automatics are not as efficient as manual transmissions, nor are they designed to be run with a motor that can be idled down to zero rpm. Modern automatics are also complicated devices that often require electronic control of shift points.

Compared to automatics, manual transmissions are much easier to implement in an EV project, are very cost effective, and can be driven by people who don't even know how to operate a manual. This latter trait is possible because, once a gear is selected, an electric vehicle can be driven to a

Electric motors are heavy and, while it's possible to rely solely on the mounting interface at the front of the motor to the transmission, it's often preferable to provide another means of support for the weight of the motor. A common means of doing this is via a motor "cradle" that is designed to transfer the loads of the motor down into the original ICE motor mounts. Shown here is a cradle designed specifically for mounting WarP and Impulse-type motors into a Chevy S-10 engine bay. Courtesy EVSource.com.

Another means of supporting the cantilevered weight of an electric motor is by way of an end plate, like the one shown here. Strong and simple, an end plate can provide the needed support for the motor for relatively little cost. Note however that, unlike a vehicle-specific cradle mount, an end plate like this will likely require some type of fabricated cross-member or attachment plate inside the engine bay. Courtesy EVSource.com.

Another simple and strong end plate bracket. Courtesy KTA Services.

complete stop without ever depressing the clutch pedal, and then started up again. Further, when used in town for moderate speed commuting, most electric passenger vehicles can get by with just one gear, which is often second or third gear. As we said earlier, an electric motor makes full torque at zero rpm, so it is often possible to just start a car in second gear and just leave it there for most in-town driving. That said, it's often useful to have other gears that can be selected. For instance, when driving on the highway, a lower gear (i.e., higher numerical) is required. Similarly, when driving up a steep hill, a higher gear (lower numerical) is needed. Also, if a transmission is used, reverse is simply a matter of selecting that gear.

So if a manual transmission is recommended, is a clutch also needed? This is an excellent question, but, unfortunately, the answer is not quite so clear cut. Any manual transmission or gearbox will allow shifting between gears when underway without use of the clutch provided the motor and transmissions are both rotating at, or near, the same rpm.

In other words, a clutch is not strictly needed. In fact, many EVs have been successfully built and operated without the use of a clutch.

That said, there are valid reasons to consider retaining a clutch in your conversion project. A clutch is useful in an EV, but not necessarily for the traditional reasons in an ICE-powered vehicle. Instead, a clutch serves the purpose of acting as a safety device that allows de-coupling of the motor from the drivetrain in the event of an emergency, such as a runaway motor controller.

Using a clutch when shifting an electric vehicle from one gear to another is also somewhat easier on the wear and tear of the internal transmission synchronizer elements. Retaining a clutch in an EV conversion project does not add very much to the overall system complexity. All of the existing hardware (clutch pedal, hydraulic lines or cable, flywheel, etc.) are already installed, so there is little work required to retain it.

If you elect to use a clutch, bear in mind that you will have to retain the flywheel. It is possible to machine, or cut down the outer diameter of the flywheel to reduce weight and the inertia of the assembly. If you do this, have the work performed by a qualified machinist who can also balance the assembly afterward. This is especially true when using an AC motor that can spin to relatively high rpm.

Some specialty EVs can be used without a transmission. The 1310 end yoke shown here slips over the output shaft of a WarP-type motor, which allows it to be mated directly to a driveshaft. If reverse rotation is required on this type of application, some type of reversing contactor arrangement is usually required to turn the motor in the opposite direction for backing up the vehicle. Courtesy EVSource.com.

An Advanced Motor Company series-wound DC motor bolted to a Porsche transaxle via an Electro Automotive clutched adapter plate and flywheel hub assembly. Courtesy David Oberlander.

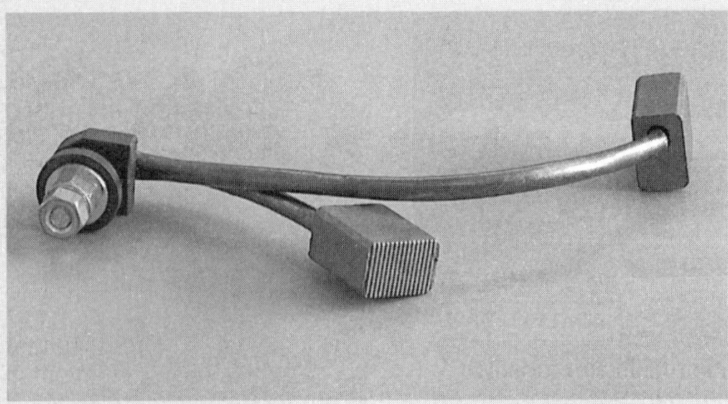

Brushed DC motors are the most common type used in EVs. They are generally less expensive than AC motors, but they also require periodic maintenance. Approximately every 20-50,000 miles of operation, the motor brushes have to be replaced. The process is fairly straightforward—provided the motor is accessible. If your electric motor is buried underneath a lot of other EV components (e.g., underneath the batteries) replacement may take a while to perform. Courtesy KTA Services.

Electric motors often work best with forced air ventilation. In addition to cooling, forced air ventilation can help minimize internal carbon dust build-up (from the brushes) that can lead to "flashovers" at high operating voltages. Shown here is a complete air-cooling kit for the popular WarP and TransWarP DC motors. Courtesy EVSource.com.

A Jabsco 12VDC motor cooling fan that is rated at 105 cubic feet per minute (cfm) of air flow. Courtesy EVSource.com.

There are two primary means of mounting an electric motor in a motorcycle. You can attach the motor to the frame or you can attach it to the rear swing arm. Neither is necessarily better or worse than the other; both have advantages and disadvantages. The next few photographs illustrate how a swing-arm mount can be accomplished. In this first image, a cardboard template is used to help decide the mounting layout on a Suzuki GSXR motorcycle rear swing arm. Once the cardboard template is finalized, it can be used to trace the pattern on a sheet of aluminum, which can then be cut out and welded to the swing arm. Courtesy Tony Helmholdt.

The aluminum bracket is welded to the rear swing arm, which has been removed from the motorcycle to allow full weld access to both sides of the bracket. Aluminum can be tricky to weld properly, so it's recommended that you hire a professional if your welding skills are poor. Courtesy Tony Helmholdt.

The swing arm is reattached to the motorcycle and the motor is test-fitted into its mounting bracket. On high-power chain drive applications like this it is important to ensure that the drive sprocket on the motor lines up perfectly and in plane with the rear drive gear on the wheel. Taking the time to measure and adjust at this point in the process is the key to trouble-free operation later. Courtesy Tony Helmholdt.

Here the rear wheel and drive gear have been reinstalled on the motorcycle, but the chain has not yet been installed. The builder of this motorcycle recommends that at this point it's important to lower the vehicle to the ground, sit on it, and bounce up and down to ensure that nothing rubs, bends, or chafes improperly. If you find a problem, fix it. Says the owner, "Once it is all done, it should look professional and well built. Remember this is a motorcycle and your life depends on your workmanship." Courtesy Tony Helmholdt.

To achieve the same performance as an ICE-powered motorcycle, this electric off-road race bike is powered by two electric motors. Courtesy Bob Simpson.

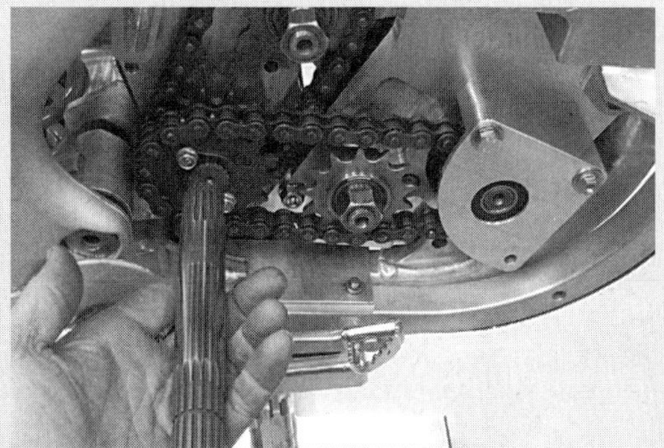

A custom two-into-one chain coupling system was designed and built by the owner. Courtesy Bob Simpson.

A motorcycle hub motor is the ultimate in simplicity. Shown here is a 10 kW motor ready to be installed into a motorcycle. Courtesy Enertrac Corporation.

An Enertrac hub motor installed on the rear swing arm of a lightweight motorcycle. Courtesy Enertrac Corporation.

A motorcycle fitted with a hub motor in action. Who says EVs are slow? Courtesy Enertrac Corporation.

Chapter 5

Motor Controllers

"Control the throttle, and you control the car."
—Juan Manuel Fangio, five-time Formula 1 World Champion

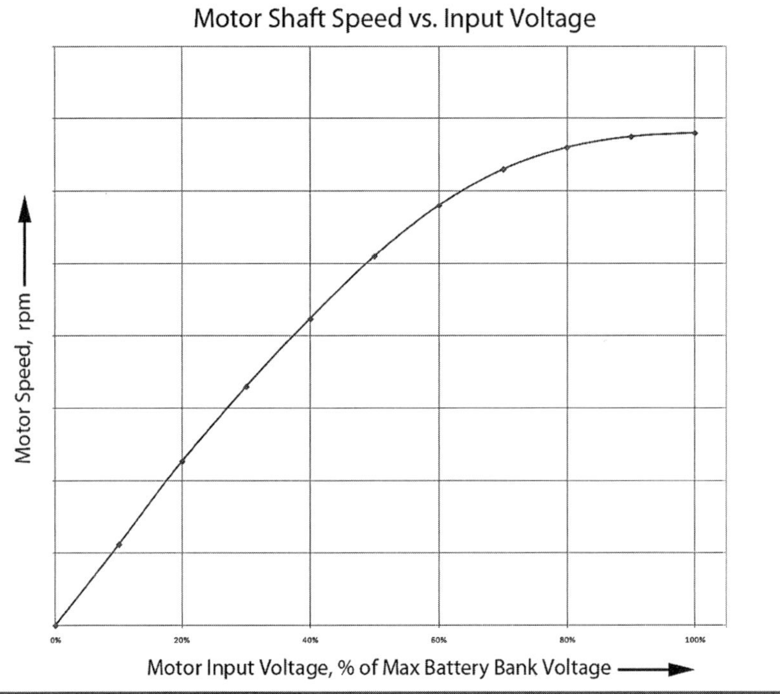

The speed of a DC electric motor is a function of the voltage that is applied to it. If we supply 100 volts to a brushed DC motor, it would spin at a fixed speed. If we cut the supplied voltage in half however the motor would slow until it was at approximately half the speed it was when 100 volts was applied. Most motor speed controllers take advantage of this phenomena and control motor speed by simply adjusting the voltage supplied from the battery bank to the motor. (In the graph, above, the roll off of motor speed at the upper end of the plot is due to back EMF of the motor; i.e., as the speed increases, the motor begins to act like a generator, creating its own back voltage, which tends to resist against the input voltage. The roll off of speed signifies the max practical output shaft speed of the motor for this particular input voltage.)

Unless you are planning to only drag race your electric vehicle and need only to switch the electric motor from fully off to fully on, you will need a smooth and reliable means of speeding up and slowing down the vehicle. Driving in town, accelerating and decelerating, merging with traffic on the highway—all of these operations require the driver to have precise control over the speed of the EV. The means by which this is achieved in an electric vehicle is via an electric motor controller, which is typically an electronic device that varies the speed of the motor based on input from the throttle, or accelerator pedal. The throttle pedal is usually attached to a device called a *pot box*, which is another electrical device that converts the position of the throttle into an electrical signal that the controller uses as a reference to vary the motor speed.

There are a number of different motor controllers available to the EV enthusiast, ranging from rudimentary low-power DC units to sophisticated, high-power AC controllers that include regenerative braking and other advanced features.

In this chapter, we take a brief look at how modern electronic motor controllers operate. We will also examine the different types of controllers available, including a discussion of their relative merits. Let's get started with a brief understanding of how the speed of a motor can be electronically varied.

Controller Basics

The job of a controller is first and foremost to allow the operator to adjust the speed of the motor. But how is this achieved? Answer: voltage.

The torque output of a DC motor is proportional to the electric current supplied to its windings. The speed of a motor, however, is a function of the voltage that is applied. If we supply 100 volts to a brushed DC motor, it would spin at a fixed speed (say, 100 rpm). If we cut the supplied voltage in half however (i.e., 50 volts), the motor would slow until it was approximately half the speed it was when 100 volts was applied (i.e., 50 rpm). Most motor controllers take advantage of this phenomena and control motor speed by simply adjusting the voltage supplied from the battery bank to the motor.

Rheostat Type—So how can we adjust the voltage at the motor input? Some early, crude controllers simply employed large, variable resistance devices, such as *rheostats* and *potentiometers*, to vary the voltage that the motor receives. The chief advantage of these systems is their simplicity; the potentiometer can be installed inline (i.e., in series) between the battery and the motor. As the resistance is changed in the variable resistor, proportionally more or less voltage is available at the motor. The chief disadvantage of this type of system, however, is its extremely poor efficiency. At low commanded speeds, the required resistance in the rheostat is very high, and, because resistors dissipate energy, the stored energy in the battery bank is essentially wasted as heat. In a typical electric vehicle, where energy storage is at a premium, this type of system makes little sense; throwing away a significant fraction of your stored battery bank energy as waste heat is dumb.

Switch Type—Another type of simple controller that has been used in the past is a *switch*, or *contactor*, controller. This type of controller uses a number of high-power contactors to connect the batteries in various combinations of series and parallel to achieve different discrete voltage settings. For example, two 12-volt batteries can be connected in parallel (12 volts) or series (24 volts) with a set of contactors for a low and high speed output, respectively. More contactors can be used on larger battery banks for progressively more discrete speeds. This type of controller isn't available for sale, per se, on the market; instead, to create a switch controller you would have to purchase the requisite contractors and then wire it up yourself. The advantages of switch controllers are they are relatively inexpensive and simple. They are also very efficient, with essentially zero voltage losses during operation. The disadvantage of course is the fixed step-size control of the motor speed, which can make driving in stop-and-go traffic difficult.

Electronic Type—While the aforementioned rheostat- and switch-controllers have their occasional places in specialty applications, the vast majority of EVs use modern electronic controllers. The most common of these are based on an electronic technique called *pulse-width modulation*, or PWM.

In simple terms, pulse-width modulation is a method of varying a voltage by rapidly switching the voltage on and off (it also can be used to vary other parameters, such as power, current, signal intensity, and so on). To illustrate how PWM works, let's imagine a 100-watt household light bulb connected to a standard wall switch. When the switch is in the off position, no electricity flows to the bulb and it stays off, using zero watts of power. When the switch in the on position, the bulb receives full power, is fully illuminated, and it dissipates 100 watts.

Now, let's imagine that instead of placing the switch in either the discrete off or on position, we instead flick it from off to on, and then back again, over and over very rapidly. The light bulb filament heats up when the switch is temporarily in the on position, but then can't fully cool down when the switch is temporarily turned off because it is turned right back on. Rapidly switching the bulb on and off like this makes the average power dissipation the bulb experiences somewhere between 0 and 100 watts. In other words, if we timed the rapid on-off switching correctly, we could create a system in which the 100-watt light bulb behaved like a less powerful 50-watt bulb, putting out a less bright glow and consuming half the energy of a 100-watt bulb. If we kept the switch in the off position proportionally longer than in the on position during the rapid cycling of the switch, the bulb would act like a lower wattage unit. Similarly, if we kept the switch on longer than off during the switching process, the bulb wattage output would be higher. Said another way, if we "modulate" the "pulse

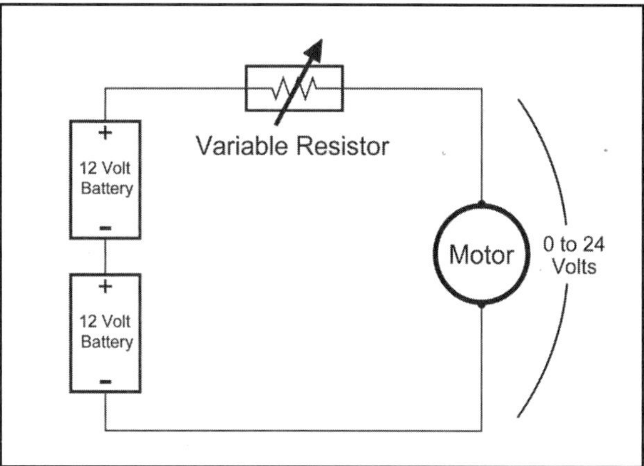

One of the simplest motor speed controllers is the variable resistance type. The chief advantage of these systems is their simplicity. As resistance is varied in the resistor, the voltage at the motor is varied by changing the resistance of the variable resistor. The disadvantage is the extreme inefficiency of the circuit. At low speeds, much of the battery energy is thrown away as waste heat.

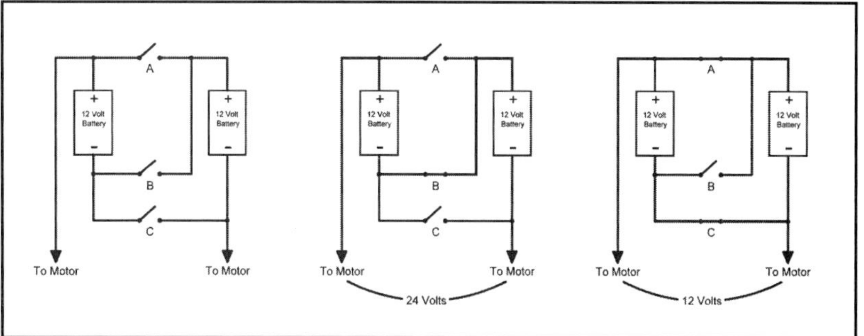

Another simple type of motor speed controller is a switch-, or contactor-controller. A number of high-power contactors are used to connect the batteries in various combinations of series and parallel to achieve different discrete voltage settings. In the example shown here, three switches are wired into the circuit. If switches A and C are left open, and switch B is closed, the output voltage of the two batteries is added in series to 24 volts. This would be the "high speed" configuration. If B is opened, however, and A and C closed, the result is a "low speed" configuration of 12 volts delivered to the battery. More and more switches can be added to battery banks that have more than just two batteries, which can allow various other combinations of parallel and series configurations, thereby allowing for a larger number of discrete motor speeds.

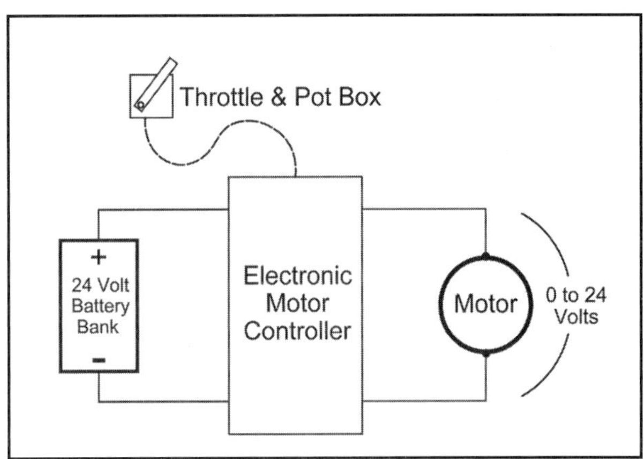

The majority of battery electric vehicles on the road today utilize a modern electronic motor controller to vary the speed of the vehicle. Most controllers of this type utilize pulse width modulation (PWM) to effectively change the input voltage to the motor. The operator controls the unit by way of the throttle pedal, which is connected to a small potentiometer, or pot box. Changes in the resistance of the pot box are interpreted by the motor controller as commands to vary the output voltage, thereby affecting the speed of the motor.

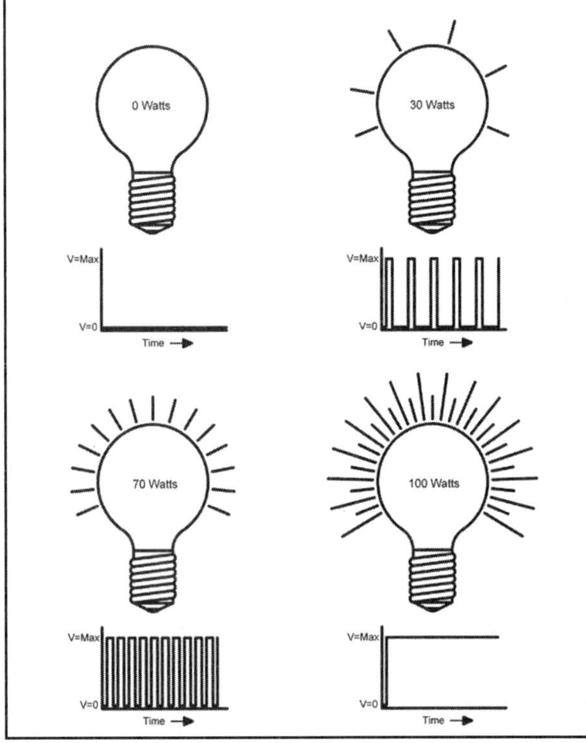

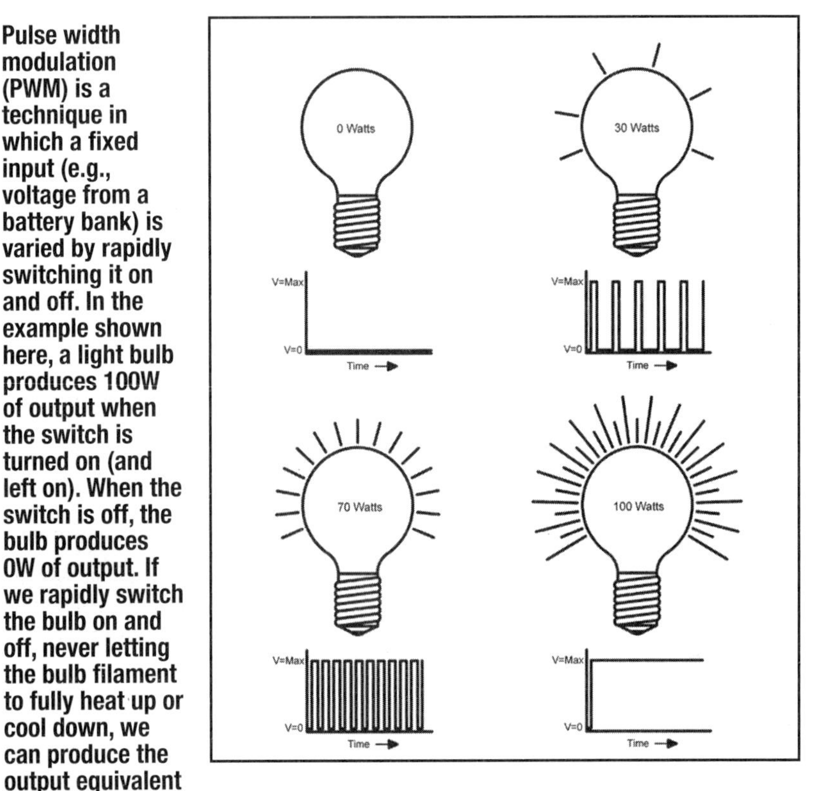

Pulse width modulation (PWM) is a technique in which a fixed input (e.g., voltage from a battery bank) is varied by rapidly switching it on and off. In the example shown here, a light bulb produces 100W of output when the switch is turned on (and left on). When the switch is off, the bulb produces 0W of output. If we rapidly switch the bulb on and off, never letting the bulb filament to fully heat up or cool down, we can produce the output equivalent of a lower wattage light bulb. The higher the ratio of "on" periods to "off" results in higher wattage output. This technique is called pulse width modulation because we "modulate" (or vary) the amount of time that the voltage "pulse" is on.

width" of the on period of the switch, we can directly affect the output wattage of the light bulb.

This same PWM technique can be applied to an electric motor. If we apply, say, 24 volts to an electric motor when a switch is closed, the motor will spin at a particular speed, say 100 rpm. If we turn the voltage off, the motor will begin slowing and will eventually coast to a stop due to bearing and other sources of friction. This coasting will take a finite amount of time to stop because the spinning rotor has mass and inertia. (This is analogous to the light bulb filament taking time to cool down when the power is switched off.)

Now, let's imagine that we apply the 24 volts to the motor for a brief instant, and then remove the voltage completely, and then reapply it, over and over in a very fast manner. By adjusting the on time versus the off time that the voltage "pulse" is applied in each one of these cycles, we can effectively change the average voltage that the motor sees. In other words, we can take a fixed voltage from the battery bank and modulate it down to some lower value by simply switching it on and off at the correct rate. This in turn allows us to control the speed of a motor from zero up to its maximum velocity. Voilá, we've invented the PWM motor speed controller.

Scientists and engineers have known about this PWM technique for many years, but it took the advent of modern electronics to apply the method to mainstream consumer products like motor controllers. Most PWM-based motor controllers use silicon controlled rectifiers (SCRs), metal oxide semi-conductor field effect transistors (MOSFETs), or insulated gate bipolar transistors (IGBTs) to perform the high-speed internal switching required for PWM operation. Each of these devices has its own pros and cons, but for the average EV builder the differences are minor and, as long as the controller is designed to handle the voltage and current levels of your motor application, it doesn't really matter which is incorporated into the design of the controller you select.

Pot Box—Okay, so we now have a device (i.e., the controller) that can be used to vary the speed of the motor. Next we need a means of translating and transferring driver input (such as the position of the throttle pedal) to the controller so that it can respond accordingly and vary the motor speed. To do this, a small device called a *pot box* is typically used. This component incorporates a modest variable resistance potentiometer, which is connected to the accelerator pedal. The potentiometer is wired to the controller via a low-level signal wire. When the accelerator pedal is depressed, the resistance of the potentiometer is changed. This change is sensed by the controller, which proportionally varies the voltage output to the motor. Most DC motor controllers use a 0 to 5000 (0–5K) ohm pot box, though some of the newer units employ somewhat different resistance ranges. Still others use a slightly more efficient

> **Selecting Controllers**
> When selecting a controller, try to download the full specification sheet or manual from the manufacturer prior to purchasing the unit. Know what it can and can't do and how it is installed and operated before you spend any money.

inductive-type throttle position sensor.

AC vs. DC Controllers—AC motor speed controllers operate in a slightly different manner from DC controllers. To begin with, they must convert the DC voltage of the battery bank into alternating, or AC voltage. This is typically done by way of an internal inverter circuit in the controller. The electronic circuitry inside an AC controller then varies the frequency of the generated AC voltage, which results in a variation of the motor shaft speed. The means by which the frequency is varied can be achieved a number of different ways, including PWM techniques.

AC controllers are generally more expensive than DC controllers, but they can offer higher efficiency, a wider usable torque band, and built-in regenerative braking. The chief downside is cost; AC controllers are typically two to three times the cost of an equivalent DC controller. AC controllers also must be matched with a specific AC motor. It is not usually possible to use an AC controller from one manufacturer with the motor from another. DC controllers, in contrast, can often be mixed and matched with different brand motors.

Choosing a Controller

A motor controller is one of the most expensive individual components purchased as part of an EV conversion project. As a result, selecting the correct unit for a particular application is a critical undertaking. Selecting a controller too large or too small or one that is mismatched for your particular motor can be an expensive mistake. Similarly, selecting a controller that is inefficient at converting battery voltage to usable power to the motor can be costly in terms of range and performance.

Most DC motor controllers are specified in terms of the nominal input (battery) voltage range that they can handle and the maximum output (motor) current they can produce. For example, the popular Curtis 1231C-7701 controller is rated for battery packs that range from 72 volts to 120 volts, and its max current rating is 550 amps.

In this example, the controller current rating stated is the maximum, instantaneously allowed

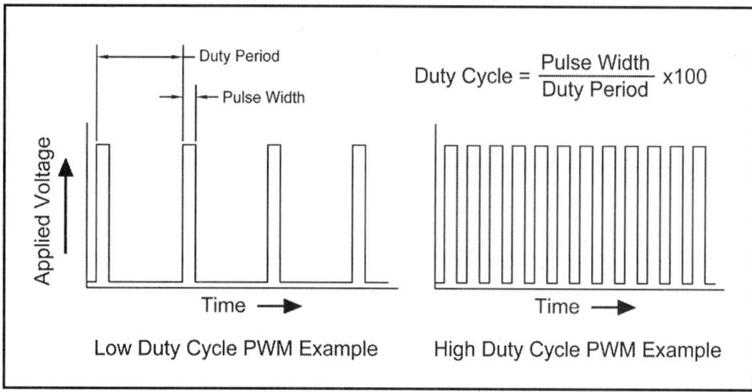

In PWM terminology, the phrase "duty cycle" is used to describe the proportion of ON time to the regular interval or duty period of time, expressed as a percent. At low duty cycles, the amount of ON time is relatively low, and therefore corresponds to a low power setting. Said another way, the power is OFF for most of the time. The output power increases as a function of the duty cycle; the larger the fraction of time the unit is ON, the higher the effective output. A duty cycle of 0% corresponds to a completely OFF system, while a duty cycle of 100% is the same as a system that is switched fully ON.

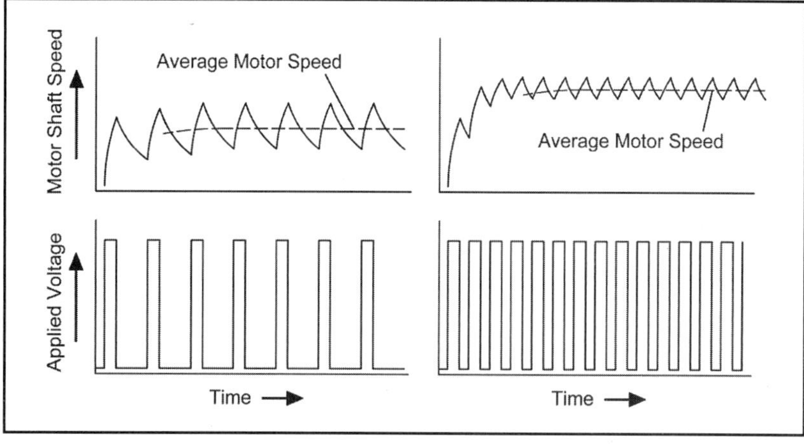

Pulse width modulation (PWM) is an effective means of controlling the speed of a motor from a fixed voltage source, such as a battery bank. This makes PWM an ideal means of controlling the speed of an EV. Most modern motor speed controllers use some type of PWM.

number of amperes that the unit can safely provide to the motor. Most manufacturers use this maximum rating in advertisements and catalogs, but it should not be necessarily relied upon when sizing the controller for continuous duty. The Curtis controller, for instance, has a max current rating of 550 amps, but this is possible only for about two minutes or less of continuous use. The unit can handle up to 375 amps for about five minutes, but if you want to run the unit in a sustained mode, the max current rating is just 225 amps for one hour of operation. For most street applications, this is more than enough, but if you're

Two of the most popular Curtis controllers are the 1221C-7401, which is capable of supporting between 72-120V battery voltages and 400 amps of current, and the 1231C-8601, which is shown here. The 1231C-8601 can support battery banks from 96 to 144V and motor currents of up to 500 amps. Courtesy EVSource.com.

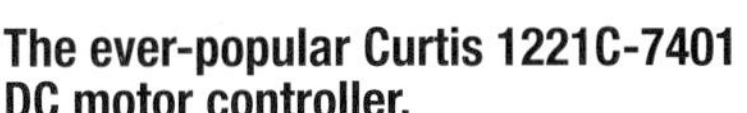

The ever-popular Curtis 1221C-7401 DC motor controller.

Alltrax manufactures a variety of quality moderate-voltage controllers, such as the DCX model shown here, which is intended for DC shunt-wired motors.

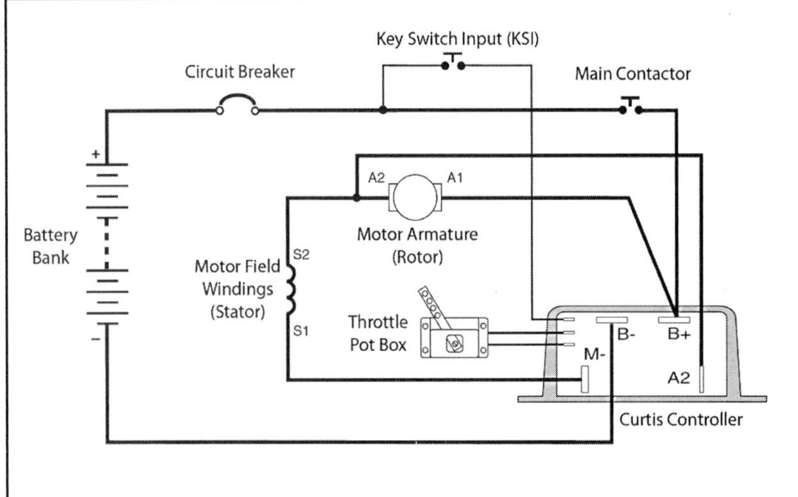

Installing and wiring a modern PWM-type speed controller is relatively simple. This schematic shows a typical wiring layout for a popular Curtis motor controller. Note that the motor is a series-wound type, with the armature (rotor) and field (stator) windings wired in series. Also note the key switch input (KSI) which is used to turn on the controller. A simple throttle pot box is connected to the controller, and used to vary the output voltage to the motor windings, thereby controlling the output shaft speed.

unsure, talk with the manufacturer.

Besides voltage and current, there are other features to consider when shopping for a DC controller. For example, what pot box is the controller compatible with? Does it come with a pot box? Does the controller have any kind of over-current and/or over-heating protection built in? What about over-voltage? And, just as important, how well does the unit handle an under-volt situation?

Another consideration is whether the controller allows for motor reversing. If not, does motor reversal need to take place via a set of external reversing contactors? Note that for many street applications, where the original vehicle transmission is retained, the ability to reverse the motor rotation direction is not necessarily required; i.e., the original transmission most likely has an operable reverse gear.

How efficient the controller is may also sway your decision. Some controllers are very efficient at converting battery energy into useable work performed by the motor, while others aren't so good. Even a few percentage point differences between two controllers can have a significant effect on range.

Most controllers have built-in fins or other devices to help dissipate excess heat, but many of these still require some kind of external heat sink, such as a large, thick aluminum plate to which the unit must be mounted. The space requirements for this heat sink should be taken into account when selecting a unit.

Another consideration is how adjustable or "tunable" the controller is. Some lower priced units do not allow the user to make any fine adjustments to the operating characteristics, while other, more expensive units do. For example, the afore-mentioned Curtis 1231C-7701 controller has adjustment pots for setting the max allowable acceleration rate, current limits, and other parameters.

There are a number of reputable manufacturers of quality motor controllers available to the enthusiast. These include Alltrax, Café Electric, Curtis, and Zapi. Choosing between the myriad of options and models can be daunting, so it is suggested that you spend time discussing your specific requirements with the manufacturer before purchasing. It's also useful to get input from other enthusiasts who are using the same unit in their EV. Sometimes the choice between controllers boils down to what units are popular in your neck of the woods. Also, it's often useful to arrange for a test ride in a vehicle that has the controller and motor you're considering installed and running. Some controllers may have all

In addition to near-zero emissions and low dollars-per-mile operation, one of the nice side benefits of electric vehicles is how quiet they are in operation. With no cacophony of under-the-hood ICE noises or loud exhaust output from the tailpipe, most EVs create very little sound when driven. Unfortunately, "most" EVs is not the same as "all" EVs. Some electric vehicles in fact can be very noisy, exhibiting something called whining.

Whining is a high-pitched squeal or chirping noise that some controllers exhibit. The problem is mostly relegated to older controllers, but some modern, low-price motor controllers also exhibit a characteristic whine under certain conditions, such as when a vehicle is traveling at low speeds and the motor is drawing high current. The whine is usually caused by a shift to lower PWM frequencies when the throttle is moved to a near-off position. The lower PWM frequency is, unfortunately, right in the middle of the human audible range. It's annoying, but perfectly normal for some controllers.

the features you require, but in the real world don't work quite as well as advertised. For example, some popular units have reputations for noisy operation, emitting loud whining or buzzing sounds when in use under certain conditions. This is especially true of older and/or off-brand units. A controller is an expensive component, and it will pay to take your time to select the best possible solution given your own budget and operational needs.

Regenerative Braking Control

Any object traveling at speed has kinetic energy. This kinetic energy is directly proportional to the mass of the object and, more important, proportional to the speed of the object squared. Double the weight of a moving car, for instance, and you double its kinetic energy. Double the speed, however, and you quadruple its kinetic energy. A 3000-lb car at 60 mph has nearly 380,000 lb-ft of kinetic energy.

All of this energy has to be converted to some other form of energy when the brakes are applied and the car is brought to a stop (remember: energy can't be created or destroyed; it can only be stored or transferred to another form of energy). By way of friction, car brakes transfer the kinetic energy of the moving vehicle into heat energy that the brake rotors and drums have to dissipate into the atmosphere. In other words, all the energy associated with the movement of the car is literally wasted into the air as heat when the brakes are applied.

If we could somehow capture (and store) some or all of that energy for later use to propel an EV, we could significantly increase the vehicle's range. Alternatively, we could reduce the number of onboard batteries to achieve the original range. But how to do this? The answer is via something called *regenerative braking*.

In simple terms, regenerative braking is a method of using the movement of the car to back-drive the electric motor, thereby turning it into an electrical generator. In other words, the kinetic energy of the vehicle is scrubbed off and converted into electrical energy by letting the motor "brake" the vehicle. The electrical charge created by the motor is then directed via the controller back to the batteries and used to recharge them on the go. This motor-generator action is used in parallel with the brakes to slow the car and bring it to a stop. Some vehicles equipped with regenerative braking can see their overall range extended by upwards of 15% or more. Regenerative braking is particularly effective in hilly terrain and/or in heavy stop-and-go traffic.

The good news is that if you have a permanent-magnet DC motor and controller, regenerative braking is often very easy to incorporate. Also, almost every AC motor and controller on the market today has built-in regenerative abilities. The bad news is that if your motor is a series-wound DC unit, it will be difficult, if not impossible, to add regenerative braking. Except for a few specialty (read: very expensive) units, most series-wound controllers are not capable of providing regenerative braking. Owners of series-wound equipped vehicles who want regenerative braking are often required to add a separate, dedicated system using a purpose-built generator connected to the driveline and activated by a separate controller or via a micro-switch on the brake pedal.

An Alltrax AXE DC series-wound type controller in the process of being installed on an electric motorcycle conversion project. Courtesy Tony Helmholdt.

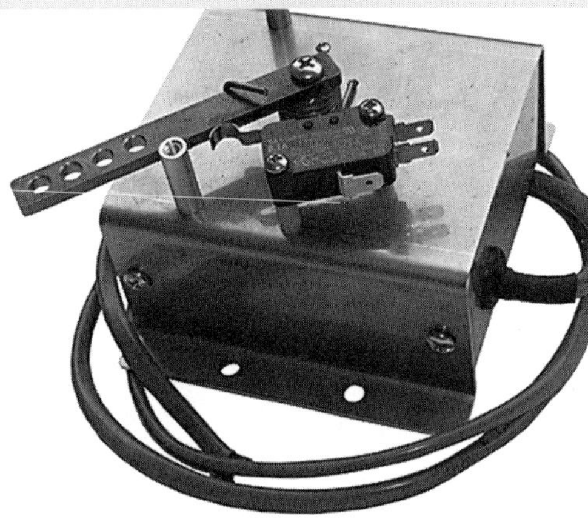

Most DC motor controllers require a potentiometer, or "pot box" to provide the input signal needed for speed control. Shown here is a typical 0-5K ohm pot pox that can be connected to the accelerator pedal cable of a car or truck. As the pedal is depressed, the resistance of the pot box changes accordingly. The controller senses this resistance change and interprets it as a command to speed the motor up. Simple and effective. Courtesy Electric Car Company of Utah.

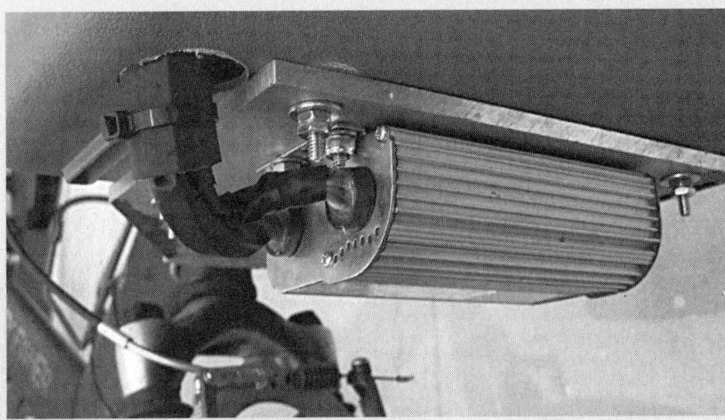

A 36-volt brushless motor controller mounted to an electric bicycle. This particular unit is routinely run at 48 volts or higher, which voids the controller's warranty, but guarantees a larger EV grin on the rider's face.

A simple thumb lever is used to operate the controller on an electric bike. Note the on-off button and the battery "state of charge" LED lights that are built into the unit.

A durable Logisystems 144 volt motor controller. Courtesy Electric Car Company of Utah.

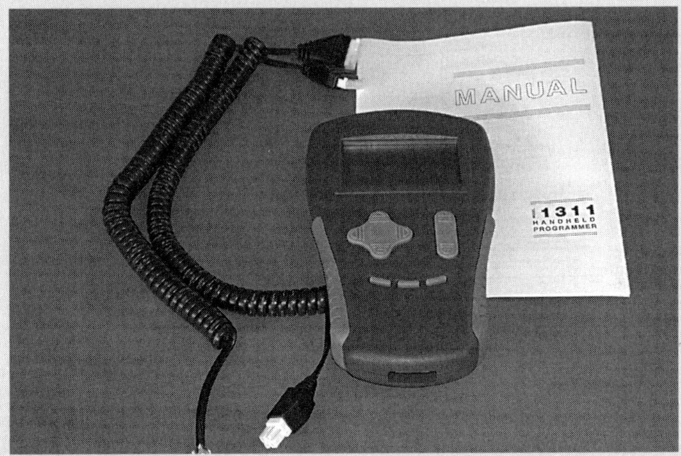

Some controllers can be reprogrammed for optimal performance. Shown here is a Curtis handheld programmer unit. Courtesy KTA Services.

This NetGain Controls WarP motor controller is both rugged and easy to install and setup. The base unit comes programmed for 1000 amps of continuous rated output at 160 volts, but can be upgraded to 1400 amps and 360-volt operation. Courtesy EVSource.com.

An Azure Dynamics/Solectria DMOC 445 II AC motor controller under the rear glass of a Bradley GT II kit car that has been converted to electric power. Courtesy Mark Bush.

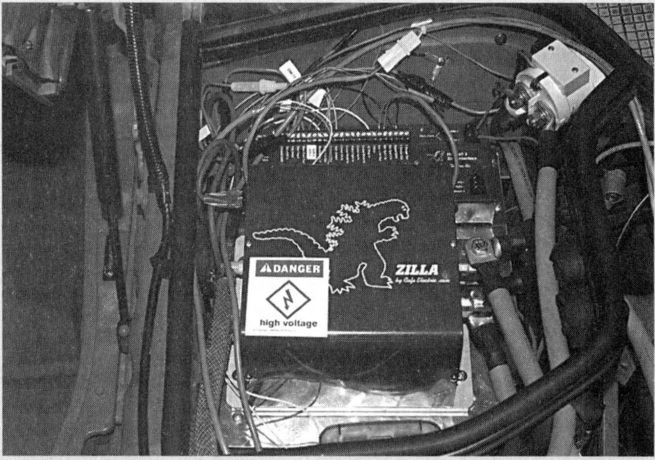

A Café Electric Zilla Z1K-HV controller installed in an electric BMW Z3 roadster. The Zilla is an excellent choice for high-performance and race-type electric vehicles. It is fully adjustable for battery voltage and current, motor voltage and current, RPM and more. Courtesy Tim Catellier.

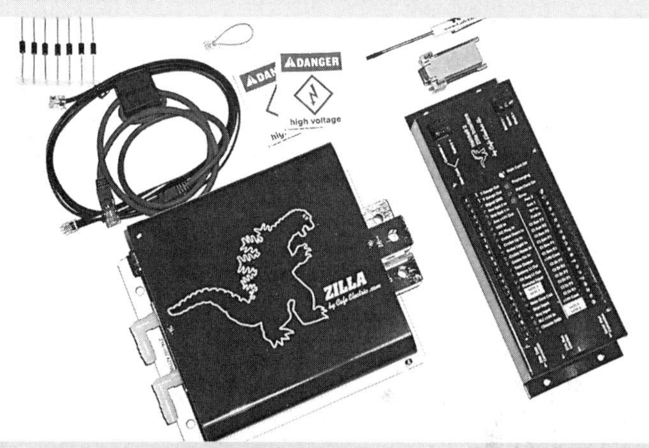

A complete Zilla controller kit, including the "hairball" interface board. Courtesy EVSource.com.

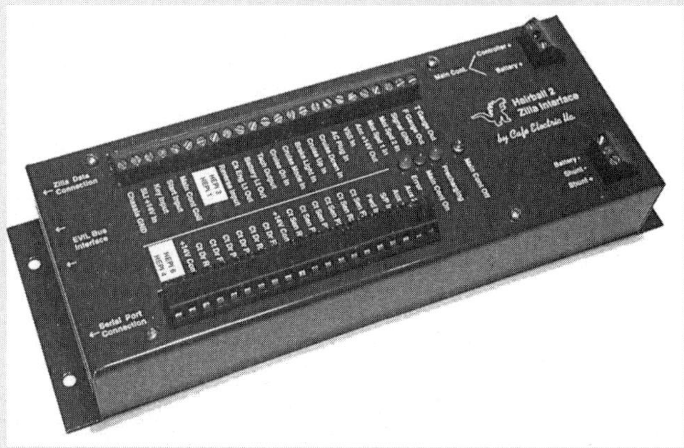

The Hairball interface board is required to operate the high-end Zilla motor controller. The board provides all input and output connections needed to operate the Zilla. Courtesy EVSource.com.

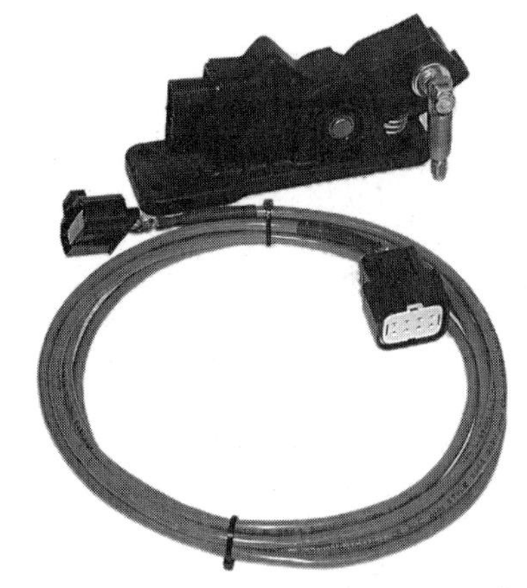

While the majority of motor controllers use a variable resistance pot box to provide the input signal for speed variations, some modern controllers are migrating toward hall-effect sensors, which are considered more efficient than a traditional pot box. Shown here is a Hall effect-type throttle pedal assembly for a WarP motor controller. Courtesy EVSource.com.

A Hall-effect pedal assembly for a Zilla motor controller. Courtesy EVSource.com.

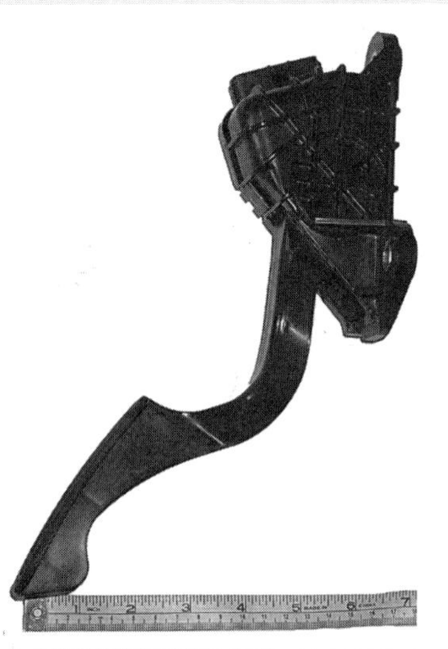

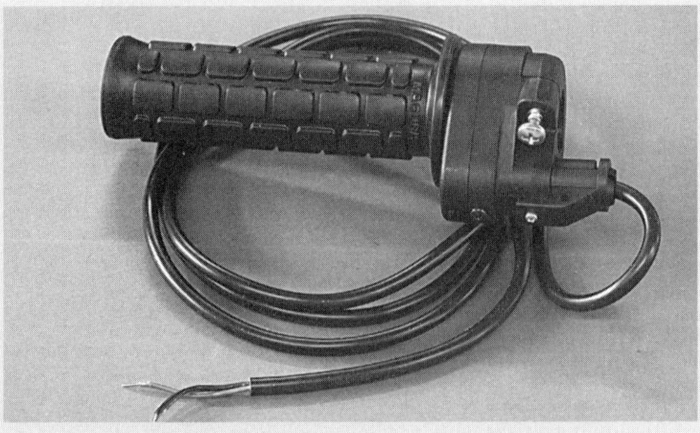

A Magura Twistgrip Throttle for use on 2- and 3-wheeled electric vehicles. The throttle control uses a built-in 0-5k ohm potentiometer for interfacing to a DC motor controller. Courtesy KTA Services.

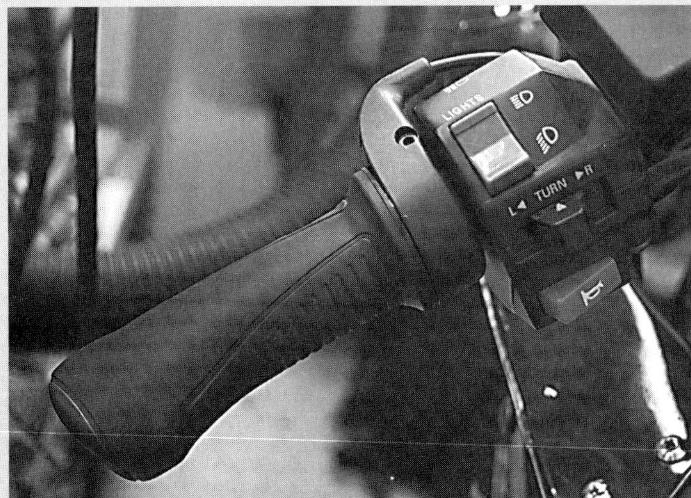

The owner of this electric motorcycle installed two Magura 0-5K twist grip throttles. One is mounted in the normal location on the right handlebar, and is used to control the output speed of the motor (via a Kelly KD72301 controller). The second twist grip throttle is installed backward on the left handlebar of the motor, as shown here. By twisting forward on the grip, the rider is able to activate and control the amount of regenerative current back fed through the controller to the battery bank. Courtesy Lennon Rodgers.

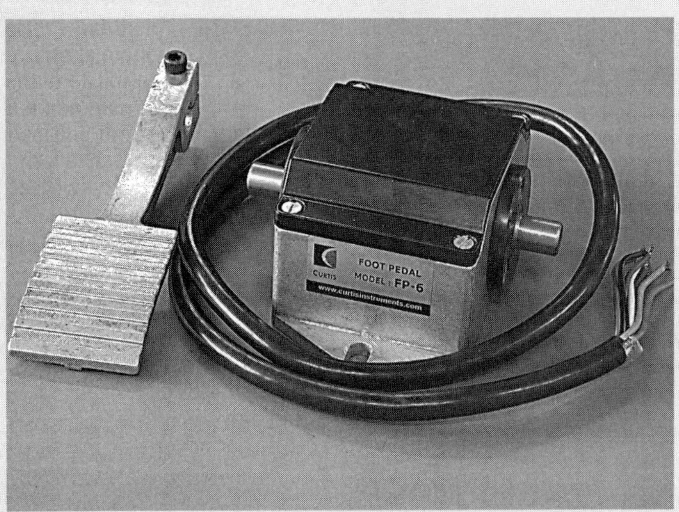

A Curtis 0-5K ohm FP-6 foot pedal unit. Courtesy KTA Services.

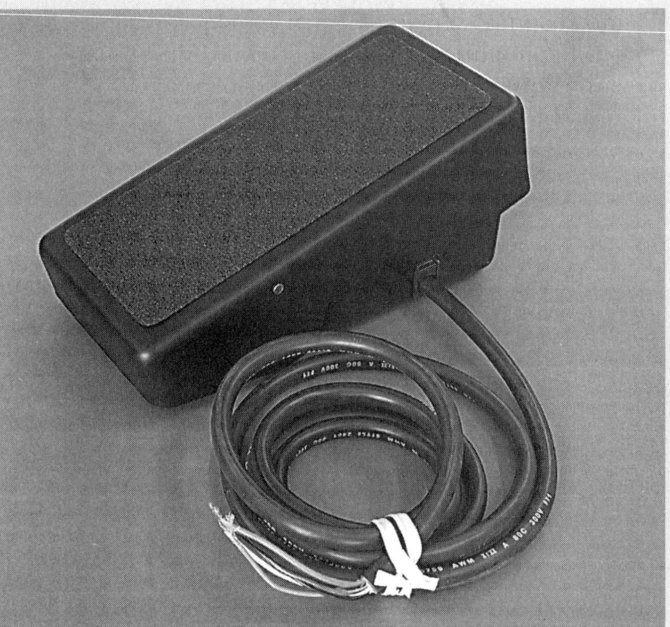

A Curtis 05-K ohm FP-2 foot pedal throttle unit. Courtesy KTA Services.

Zilla controllers are designed for water cooling. The liquid cooling system kit shown here includes a small radiator, electric pump, hoses, and the clamps required to provide heat removal from the controller. Courtesy EVSource.com.

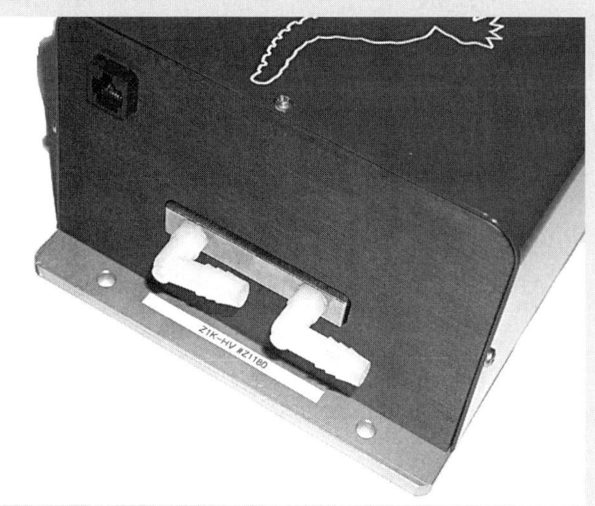

A close-up view of a Zilla controller that shows the input and output connections for the external liquid cooling system. Courtesy EVSource.com.

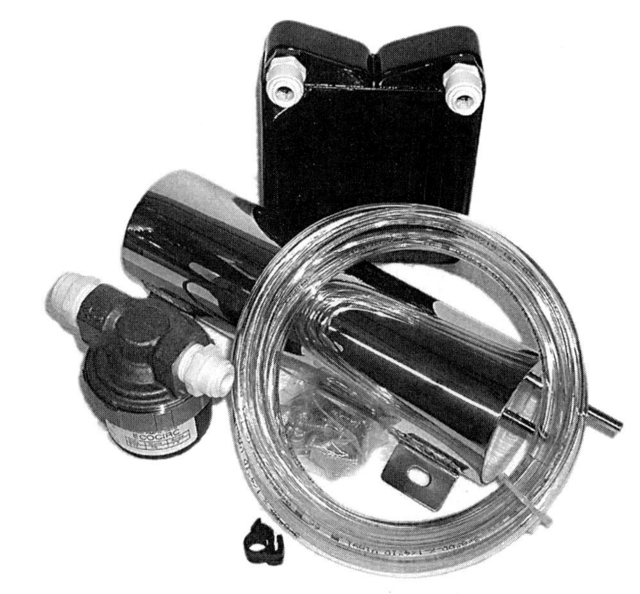

This liquid cooling system is designed for use on the WarP motor controllers. Courtesy EVSource.com.

Some controller manufacturers recommend the use of a "pre-charge" resistor that is wired across the contact terminals of the main contractor. This allows the controller's internal capacitors to charge more slowly before the contacts are switched closed. Courtesy EVSource.com.

Chapter 6
Batteries

Flooded lead-acid batteries are the workhorse of energy storage in most electric vehicle conversion projects. Simple, robust, and reliable.

"It has been said that batteries are the worst form of EV energy storage possible...except for all the others that have been tried."
—anonymous battery engineer

Perhaps the most significant hurdle that has kept electric vehicles from becoming mainstream solutions to society's transportation needs is battery technology. Batteries are heavy, expensive, slow to charge, have limited storage capacity, and need to be replaced every few years. Sounds terrible, right? Well, yes and no. While modern EV batteries leave much to be desired (especially in the area of energy density), they also offer a number of very useful benefits. For one, they're very simple devices and can be manufactured from a variety of different materials and processes. Modern batteries are also rugged and incredibly reliable. In addition, they have no moving parts and are nearly 100 percent recyclable. More important, they're the most practical energy storage device currently available for use in EVs, and advancements in materials, packaging, and chemistry continually result in incrementally better performance and lower cost every year. Like it or not, batteries are a key component of traditional electric vehicle conversions.

In this chapter, we will begin with a discussion of the concepts of energy and work, including how the former can be stored in a battery, so that it can be released to perform the latter. We will then discuss in detail how batteries actually operate, including how they can be recharged. We will then go into specific detail, describing the multitude of types and configurations of batteries available to you, the EV builder, along with a discussion of the relative pros and cons, capacities, and ratings for each type. We will then close out the discussion with some practical considerations, along with both some general and specific recommendations to help you choose the right battery for your EV. Sorting through the myriad of battery options and types doesn't have to be as onerous as it may first appear. So let's begin our discussion with a look at the basic concept of energy itself.

Energy, Work, and the Battery

Ask an engineer or physicist what *energy* is, and he or she will undoubtedly tell you it is the "ability of an object or system to do work." The word *work* is what happens when a force physically moves an object a certain distance. If we want to move, say, a 100-lb stack of bricks from the floor onto a table that is three feet high, we need to perform 100 lbs x 3 ft = 300 lb-ft of work. Said another way, we need to expend 300 lb-ft of energy to perform this amount of work. Just as it's possible to express distance with different units (e.g., inches, feet, miles), energy can be discussed in a number of units, including the aforementioned lb-ft, as well as ergs, joules, and, in the case of electric batteries, watt-hours.

Energy can take many forms. For example, there are mechanical forms of energy. These include kinetic energy, which is the energy that a moving object has by virtue of its mass and speed. Potential energy is another example, and is the amount of energy that an object with mass has when it is raised to a height. In the case of the brick example, we have transferred 300 lb-ft of potential energy to the system of bricks sitting on the table. In a sense, we have "stored" that energy for later use. For instance, if the bricks were in a basket, and a rope and pulley were attached to the handle, we could perform useful work by having gravity pull the weighted basket back down the ground. This is the same type of technique that old-fashioned grandfather and cuckoo clocks employed to keep running. Another example of potential energy is water reservoirs in the mountains outside some towns. The water is diverted through a turbine as it flows to a lower location; the turbine drives a generator and electrical energy is created.

There are other forms of energy that can be stored. For example, the chemical energy contained in a gallon of gasoline is essentially stored for later use. We can convert this chemical energy into thermal energy by burning it (say, in the combustion chamber of an internal combustion chamber) and then converting that into mechanical work by way of a piston, connecting arm, and crankshaft.

If we look at these examples, we see that one form of energy can be converted to another. The laws of physics state that energy cannot be created or destroyed, but it can be moved from one place to another, or converted from one form to another, or stored.

Electricity is another common form of energy. The movement of electrons through a conductor can be used to perform useful work, such as through the operation of an electric motor. The energy used to power an electric vehicle is stored in the batteries in the form of chemical energy. When a motor is connected to the battery, the chemical energy is converted to electrical energy, which in turn is converted to the kinetic energy of the rotating motor shaft, which is used to produce work (i.e., move the car).

When a charger is connected to a battery, we are in a sense reversing the process. Electrical energy from the power grid is converted back into chemical stored energy in the battery.

Note that there are other, more exotic ways of storing energy in a vehicle, such as fuel cells (chemical energy), supercapacitors (electrical charge energy), flywheels (mechanical energy), and even compressed air (pressure energy). For most of us converting vehicles to electric power, however, chemical batteries offer the easiest and cheapest energy storage system.

Finally, before we move on to discuss batteries, we need to understand the concept of efficiency. While it is true that energy can neither be created nor destroyed, some of it can and will be transferred out of the system whenever it is transferred from one form to another. This "loss" is due to inefficiencies in the system. For example, in our brick and pulley example, there is a small amount of friction in the pulley and rope that doesn't allow a full 100% transfer of potential energy into useful work. While there might be 300 lb-ft of stored brick energy, we might be able to capture only 290 lb-ft of that energy, with 10 lb-ft lost as heat energy due to friction. In other words, there are no perfect energy transfer systems; i.e., there are always efficiency losses when converting from one form of energy to another.

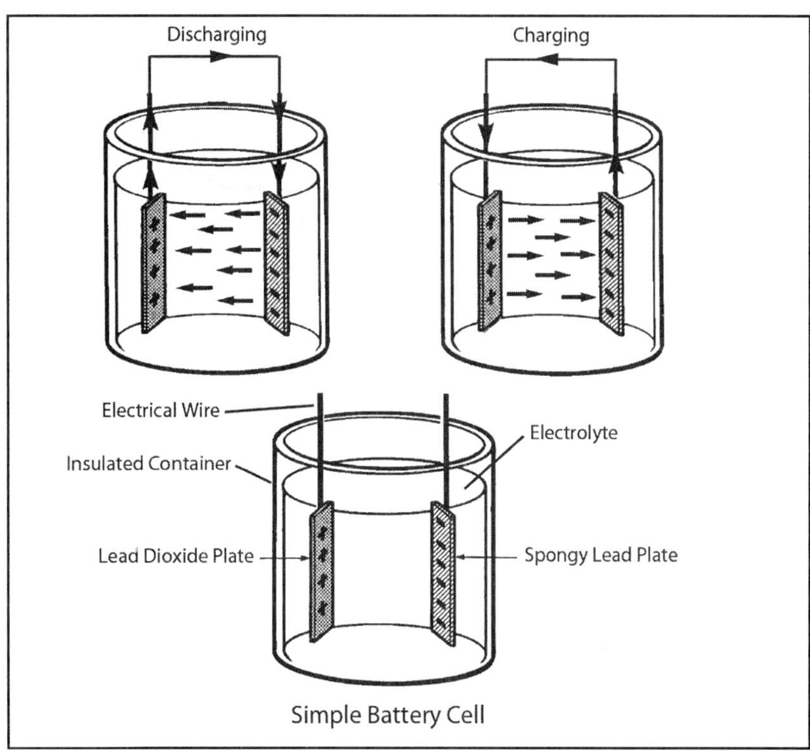

A simple battery cell can be created by filling an insulated jar with a strong sulfuric acid electrolyte and suspending two plates in the solution. One of the plates is made from lead and the other is lead dioxide. An electro-chemical reaction takes place between the lead and the acid to create a flow of electrons. All lead-acid batteries operate on this same principle. Courtesy Toyota Motor Corporation.

How a Battery Works

As most of us were all taught in high school chemistry class, atoms are the basic building blocks of matter. Atoms themselves are made up of a central nucleus of positively charged protons and electrically neutral neutrons. Atoms also contain negatively charged electrons, which orbit the nucleus and are held in place by a strong electromagnetic force.

Batteries are categorized into two major classifications: primary and secondary types. Primary batteries, like the consumer grade alkaline units shown here, are non-rechargeable. Once they are discharged, they cannot be reused without disassembly and recycling. Secondary batteries, on the other hand, are fully rechargeable. Batteries that fall into this category include lead-acid, nickel-cadmium, nickel-metal hydride, and lithium-ion types.

Chemical Reactions—When atoms are combined together, they form molecules. For example, when two atoms of oxygen and one of hydrogen are combined, the resulting molecule is H_2O, or water. Similarly, in some cases when one type of molecule comes in contact with a different type of molecule, they recombine to form a new molecule. This process is called a *chemical reaction.*

There are many types of chemical reactions, ranging from simple synthesis combinations, like the aforementioned creation of water from hydrogen and oxygen atoms, to much more complex—including displacement, decomposition, and even combustion—types of reactions. In some cases, a chemical reaction not only results in a new chemical molecule, but also causes the release of one or more of the electrons from the atoms. This latter type of reaction is the basis of how batteries operate.

To see how all this works let's look inside a standard automotive-type flooded lead-acid battery. The typical 12-volt starting, lighting, and ignition (or SLI) battery found underneath the hood of an ordinary gasoline-powered vehicle is known as a lead-acid battery. If we were to cut open one of these batteries, we would find a series of vertical metallic plates, each separated by a small distance from the next, and immersed in a liquid solution that is called an electrolyte. Some of the plates are made from castings of lead dioxide, and others are from a spongy metallic lead material. The lead dioxide plate is known as a positive plate and the spongy lead is known as the negative plate. The plates are arranged in an alternating sequence of positive and negative plates. Each pair of positive and negative plates is known as a *cell*, and the combination of cells together form the overall *battery.*

The electrolyte liquid that the plates are immersed in is a mixture of sulfuric acid and water. Sulfuric acid is an active compound of hydrogen, sulfur, and oxygen (chemical symbol: H_2SO_4). When mixed with water, the sulfuric acid molecules separate into two different types of ions, or charged molecules. These two ions are hydrogen and *sulfate*, which is a mixture of sulfur and oxygen atoms. Each sulfate ion contains two extra electrons and therefore carries a negative electrical charge. Similarly, each hydrogen ion, having been stripped of one of its electrons, carries a positive electrical charge.

The chemical reaction between the spongy lead (negative plate) and the sulfate ions produces lead sulfate on the surface of the plate. This reaction results in two free electrons which reside on the plate. Much like a north-type magnet repels another north magnet, the negatively charged electrons repel other negatively charged sulfate ions. This means the reaction starts to slow down with the buildup of a negative electrical charge on the plate.

Now, on the positive plate a similar reaction occurs, except electrons are lost and a net positive charge builds up. This reaction is also self-limiting. The difference in electrical charge between two adjacent spongy lead and lead dioxide plates is known as the electromotive force (EMF), which is otherwise known as a voltage difference. In a typical automotive battery, a voltage difference of approximately 2 volts is developed between each set of plates. This EMF is known as the open circuit voltage of the cell. There can be any number of cells in a battery. Because automotive batteries typically have six sets of plates, the total voltage difference between one side of the battery and other is 6 x 2 = 12 volts.

This total 12-volt open circuit condition remains unchanged as long as there is no pathway for the buildup of electrons to escape. In this state, the chemical activity begins in earnest when the electrolyte is added to the space between the plates. As time passes, however, the activity eventually slows to a stop. This is why batteries can remain charged for a relatively long period of time without losing much energy.

Once an electrical circuit is connected to the battery, electrons are free to move, and after an electron has left a cell plate, a replacement from the electrolyte can take its place. In other words, an electrical current can flow from the negative terminal of the battery to the positive by way of a conductor.

A typical 6-volt flooded lead-acid battery is comprised of three 2-volt battery cells, like the one shown here in schematic form. A 12-volt battery would be made from six such cells.

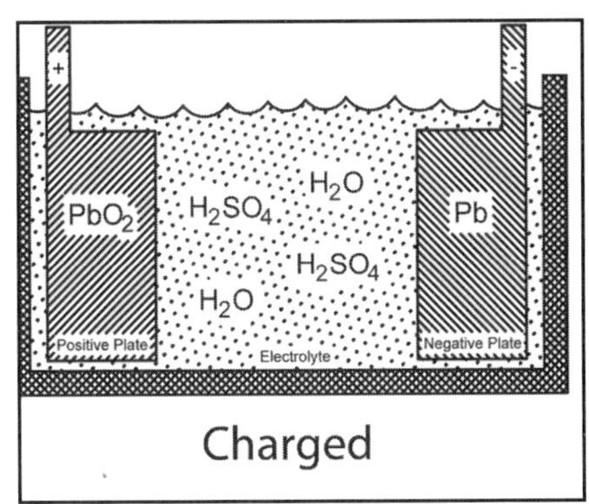

Discharging

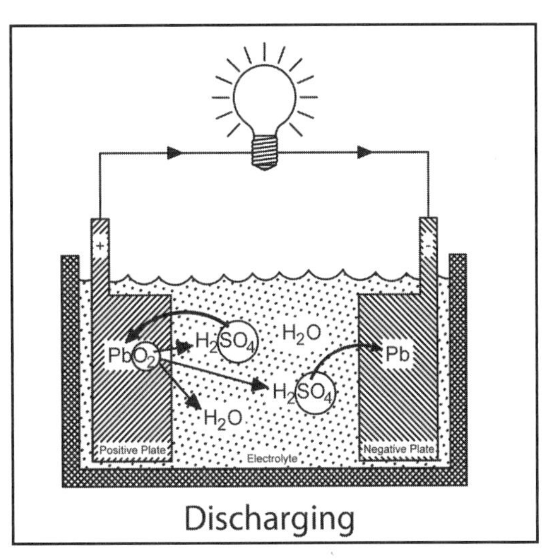

When a load is applied between the two electrode connections, an electrical current flows. The source of this current is electrons that are released when sulfate (S04) from the electrolyte combines with lead from lead sulfate (PbS04). Lead dioxide, hydrogen ions, and S04 ions, plus electrons from the lead plate, create PbS04 and water on the lead dioxide plate. As the battery discharges, both plates build up PbS04 and water builds up in the acid. The characteristic voltage is about 2 volts per cell, so by combining six cells you get a 12-volt battery.

Battery Types

Batteries are sold in an extremely broad variety of shapes and sizes. They're also available in a wide range of voltages, capacities, and chemistries. While daunting at first, understanding the differences between the different types of batteries is vitally important. Knowing what the pros and cons of each type are before selecting a battery type will make your job of converting a car to electric much more enjoyable.

All batteries, regardless of chemistry, can be classified into one of two major groups: primary or secondary. Primary batteries are also known as *standard*, or non-rechargeable, batteries. The internal chemical reaction that supplies the electrical current in a primary battery is irreversible. Primary batteries are cheap, readily available in most corner grocery stores, and are used to power standard household goods, such as flashlights, watches, toys, and radios. Most AA, C and D-size batteries are primary types. Some of these are known as *zinc* batteries, and incorporate an internal arrangement of zinc and carbon plates with an acidic paste between them that serves as the electrolyte. Others are known as *alkaline* batteries and use zinc and manganese-oxide electrodes separated by an alkaline electrolyte. Many of the popular consumer batteries manufactured by Duracell and Energizer are alkaline-types. Because primary batteries are non-rechargeable, they obviously are not a good choice for electric vehicles.

In contrast, secondary-type batteries are fully rechargeable. Secondary batteries are also known as *storage-type* batteries because energy from an outside source can be added to and stored in this type of unit for later use. Secondary batteries are used in a variety of commercial and specialty applications, including starting, lighting, and ignition (SLI) in automobiles; uninterruptible power supplies (UPSs) for emergency and backup power; telecommunications, portable cordless-type power tools; and, of course, electric vehicles. When used in an EV, secondary batteries are often referred to as *traction-type* batteries.

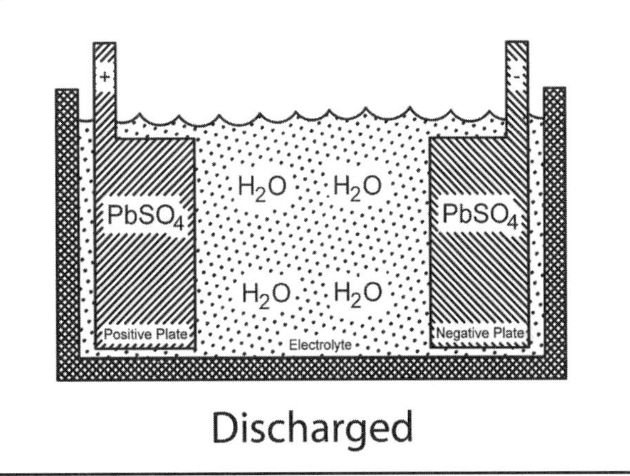

A battery cell shown in a discharged state. Note that the electrolyte is mostly water at this point, and that both plates are sulfated.

Secondary batteries can be manufactured from a variety of different materials and via a number of different techniques and packaging arrangements. These include lead-acid, lithium-, nickel-, and zinc-based batteries. Let's look at each of these in a little more detail.

Lead-Acid—By far, the most common secondary-type battery used in electric vehicle conversions is the ubiquitous lead-acid battery. As we saw above, lead-acid batteries utilize alternating internal electrode plates of lead and lead-oxide with a sulfuric acid electrolyte in between. The resulting chemical reaction with the acid and the electrode plates frees electrons, which are used to provide an electrical current. When an external current at the correct voltage is applied to the battery itself, the chemical reaction inside the battery reverses. This causes a regeneration of the lead and lead-oxide, thereby recharging the battery. In a sense, outside

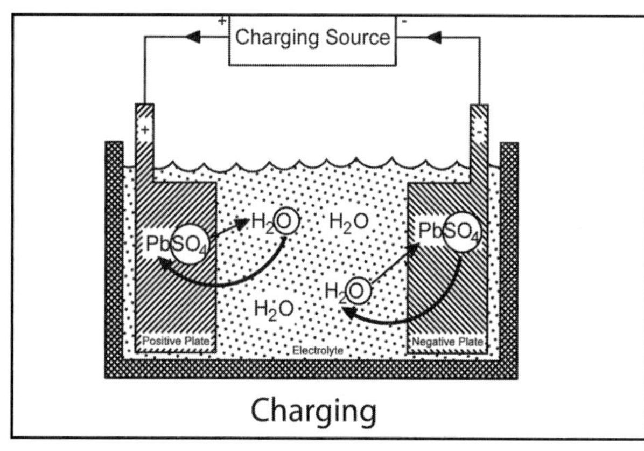

Fortunately for EV enthusiasts, lead-acid battery cells have a nice feature—the discharge reaction is completely reversible. If you apply reverse current to the battery at the right voltage, lead and lead dioxide will form again on the plates so you can reuse the battery over and over. Battery cells like this are called *secondary* or *rechargeable* cells.

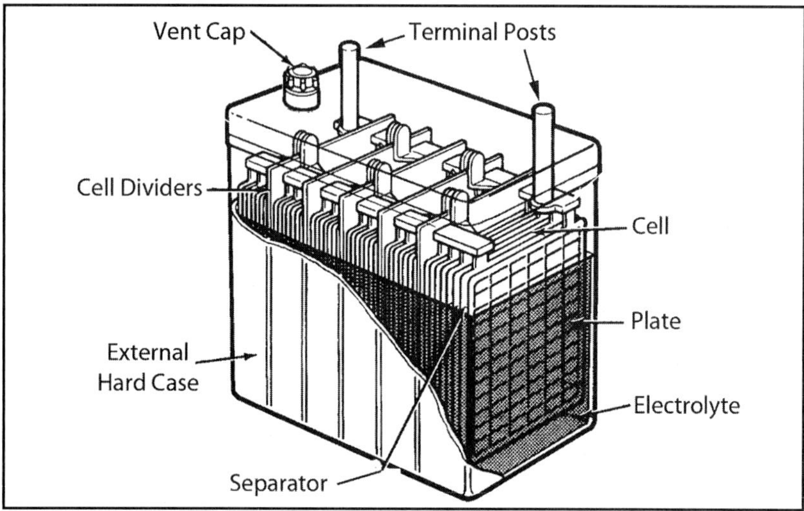

A cutaway view of a typical flooded 12-volt lead-acid battery. Courtesy Toyota Motor Corporation.

electrical energy can be added to the battery, where it is converted to chemical energy and stored for later use.

Lead-acid batteries can be categorized by both their construction and their usage. For example, lead-acid batteries are typically divided into three broad categories: (1) starting, lighting, and ignition (SLI); (2) standby power; and (3) motive, or traction types. All three types use lead and lead-oxide plates immersed in an acidic electrolyte, but their internal geometries and sizes vary. SLI batteries, for instance, have relatively thin electrode plates and are designed to provide high current discharges for relatively short periods of time (e.g., for engine starting). They are not designed to be deeply discharged, but instead are configured to provide thousands of cycles of shallow discharges (typically less than 20% depth of discharge) that are followed immediately by recharging via an alternator or generator.

Standby power batteries are used to provide energy when the main power system of a device or building is interrupted, such as during power outages. The primary application areas include telecommunications, uninterruptible power systems (UPSs), switchgear, emergency lighting, and security applications. These types of batteries spend much of their life fully charged and in a holding mode. The charging system used for these batteries maintains what is known as a *float* charge, with the batteries experiencing only very shallow discharge, except on the rare occasion that they are used, which is often to the point of full discharge. Standby power batteries are generally much heavier and bulkier than a comparable SLI or traction battery, but this is not usually a problem as they are typically used in stationary (i.e., non-vehicle) applications.

The third category of lead-acid batteries is the one most EV builders are most interested in: namely, traction batteries. In contrast to SLI batteries, traction batteries do not typically need to discharge an extremely large amount of energy in a short period of time (e.g., for powering a starter motor). Instead, a traction battery usually needs to maintain a moderate discharge rate for a relatively long period of time and be used down to a fairly deep depth of discharge before requiring recharging. To accomplish this, traction batteries typically have much thicker internal lead plates and higher energy density designs. Unfortunately, light weight is often a priority in electric vehicles, which is not something that the thick internal lead plates naturally lend themselves to.

Okay, now that we have a grasp of the major categories of lead-acid batteries, let's look at the different ways in which they can be constructed. Typically, lead-acid batteries are built in three different varieties: (1) flooded type; (2) gel-type; and (3) absorbed glass mat (AGM).

Flooded Lead-Acid—The most common type of lead-acid battery is the *flooded type*, which is also known as a *wet-cell battery*. Flooded batteries employ a liquid electrolyte and often have small access ports (with caps) that can be used to inspect, test, and replenish the acid. Flooded-type batteries are typically the least expensive variant of lead-acid batteries. They are also rugged, reliable if maintained properly, and widely available from many different sources and suppliers. They are also made in a wide range of sizes and voltages to fit almost every need and niche. There are, however, a few drawbacks to consider.

The first major disadvantage with flooded-type

U.S. Battery manufactures a line of quality deep-cycle lead-acid batteries that are popular with EV constructors. Shown here is a flooded 12-volt unit. Courtesy U.S. Battery.

Another deep-cycle battery from U.S. Battery. Note the extra tall design, which is due to tall lead plates inside the unit. This 145XC model has a 20-hour discharge ability of over 250 amp-hours. Courtesy U.S. Battery.

Deka flooded deep-cycle batteries are popular with owners of low speed, high-range EVs such as golf carts and neighborhood electric vehicles. Courtesy East Penn Manufacturing.

batteries is that they require periodic maintenance, including testing of the electrolyte and replenishing it as required with distilled water. When being charged, hydrogen and oxygen gas are generated, which comes at the expense of water. There are some specialty hydro caps that help facilitate the recombination of the oxygen and hydrogen back into water, but these are only considered a means of slowing the process and do not take the place of regular battery testing and topping off.

There are so-called maintenance-free flooded batteries available that have non-removable caps, but most knowledgeable builders of EVs generally avoid this type, mainly due to longevity concerns; a common joke among EV owners is that maintenance-free is the battery manufacturer's code for "will die one week after the warranty runs out." This is a particular problem if the battery is overcharged too many times. Sealed batteries are not usually totally sealed as they must allow gasses to vent during charging. Because of this, enough water can eventually be lost due to evaporation and out-gassing that maintenance-free batteries can die long before an equivalent open-cell design would.

Another problem with flooded batteries is due to the nature of their construction and means of supporting the lead plates. Flooded cell batteries are mechanically the weakest type of battery because each lead plate is supported only along its edges (i.e., to allow electrolyte to flow freely between the plates). For high-vibration and extreme-duty applications, alternate battery constructions should be considered.

Finally, perhaps the biggest issue with flooded cell batteries is one of leakage and spills. The sulfuric acid electrolyte used in these types of batteries is caustic and can cause burns. Flooded batteries must be mounted securely in upright orientations and be readily accessible for the owner to easily perform the required maintenance.

Flooded lead-acid batteries have been used in more home-built EVs than any other type. Shown here is the rear battery bank used in a custom fiberglass-bodied two-seat roadster. Courtesy Ed Matula.

Gelled Lead-Acid—In contrast to flooded lead-acid batteries, gelled batteries (also known as gel-cells) employ an electrolyte that has been gelled via the addition of silica gel. This turns the acid into a highly viscous, gelled mass that resembles Jell-O. This essentially captures the electrolyte in a solid form between the lead and lead-oxide plates.

The chief advantage of gelled batteries is that they will not spill acid if they are inverted or are involved in an accident when they break. The downside, however, is that over time the gel slowly hardens and dries out, which creates tiny cracks and fissures that reduce the effectiveness of the battery, allowing

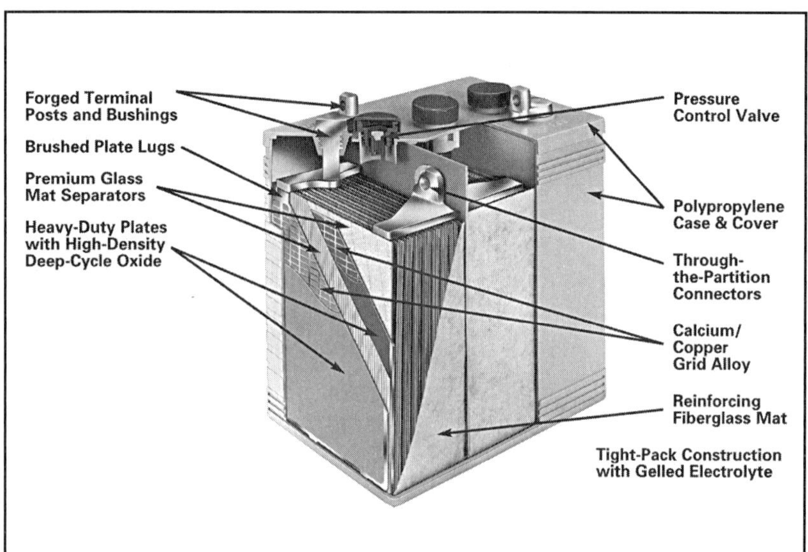

Gel-type lead-acid batteries utilize an electrolyte that has been gelled with the addition of silica gel. This turns the acid into a highly viscous, gelled mass that resembles Jell-O. This essentially captures the electrolyte in a solid form between the lead and lead-oxide plates. Courtesy East Penn Manufacturing.

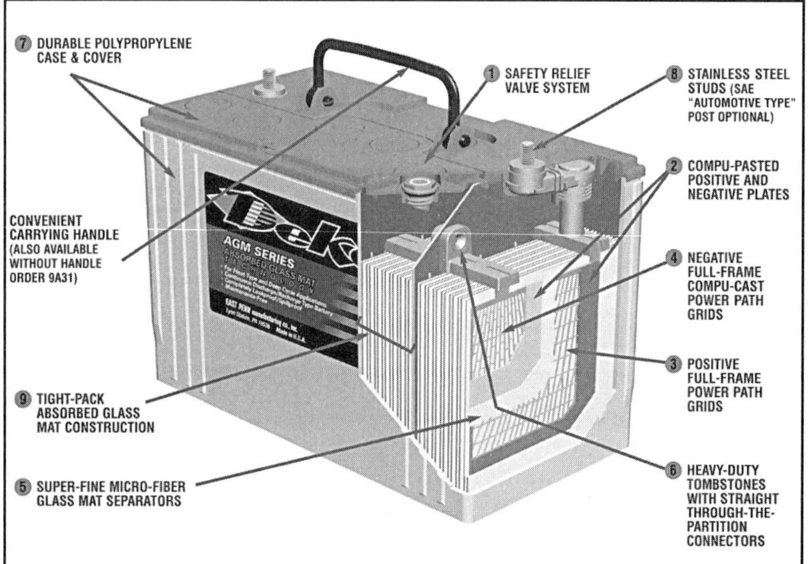

Instead of a liquid electrolyte, absorbed glass mat (AGM) lead-acid batteries utilize fiber glass mats that are placed between the lead and lead-oxide plates. These mats are then saturated with electrolyte acid, which eliminates the leakage and spillage concerns, even if the outer battery case is broken. These types of batteries can be mounted in any orientation without fear of leakage, even upside down. Courtesy East Penn Manufacturing.

oxygen to diffuse between the plates. Gelled batteries also require a relatively slow rate of recharging or else excess gas formation can damage the cells. Specialty chargers are sometimes required to limit input current to the batteries.

Another benefit of AGM-type batteries is that they can be manufactured in non-standard shapes. Optima batteries, for instance, use curved lead plates kept apart by the electrolyte-soaked fiber mat. The result is a very energy dense and structurally sound package that can withstand heavy vibration and shock loads. Courtesy Optima Batteries.

Absorbed Glass Mat Lead-Acid—To help solve the problem of gel hardening, manufacturers have recently developed an alternative called absorbed glass mat, or AGM-type batteries. Instead of a gooey gel, AGM batteries employ fine fiber boron-silicate glass mats between the lead and lead-oxide plates. These mats are 95% saturated with the electrolyte acid, which eliminates the leakage and spillage concerns, even if the outer battery case is broken. There is no gel that hardens with time, and the batteries can be mounted in any orientation without fear of leakage, even upside down.

AGM batteries are typically designed to be recombinant, which means that any oxygen and hydrogen created during charging are able to recombine inside of the battery. This keeps water loss essentially to zero, with no emissions of dangerous fumes. They can also have very low internal resistance, which means that they can be charged and discharged as quickly as the best flooded cell batteries, with little heating. Standard charging voltages and methods can be used. In addition, AGMs have a very low self discharge rate, which means that they can essentially sit unused for months at a time without suffering any degradation or loss of ability to hold a charge.

Another benefit of AGM-type batteries is that they can fairly easily be manufactured in non-standard shapes. Optima batteries, for instance, use curved lead plates that are kept apart by the mat. The result is a very dense and structurally sound package that can withstand heavy vibration and shock loads.

So if AGM batteries are so good, why doesn't everyone use them? The answer is simple: cost. Absorbed glass mat batteries typically cost two to three times as much as a comparable flooded-cell battery. And, just as important, they don't necessarily last any longer when put into service in a standard electric vehicle. For normal applications,

U.S. Battery makes a line of popular AGM batteries, including the AGM-2000, shown here. Courtesy U.S. Battery.

Lithium iron phosphate, or $LiFePO_4$ batteries, are a relative newcomer to the world of EVs. While $LiFePO_4$ batteries are not quite as energy dense as other lithium-ion batteries, they are still significantly better than lead-acids and can pack three to four times as much energy per kilogram (e.g., 100-160 Wh/kg vs. 30-40 Wh/kg). They are also capable of being cycled (charged and discharged) many more times than a lead-acid battery before requiring replacement. They also can be discharged much more deeply than a lead-acid battery without affecting life or performance. The downside, of course, is that $LiFePO_4$ batteries are more expensive than lead-acid batteries. Courtesy Electric Car Company of Utah.

like truck and commuter car conversions, AGMs are probably overkill, but if your application is more high-performance in nature, such as a motorcycle or sports car, AGMs provide some nice benefits for the additional cost.

Lithium-Ion Batteries

Not to be confused with lithium batteries, which are disposable primary (i.e., non-rechargeable) units often found in small consumer products such as calculators and watches, lithium-ion batteries are a type of secondary, or rechargeable battery gaining favor among EV enthusiasts. We will discuss why this is so in a moment, but first let's understand how a lithium-ion battery is constructed.

Like most battieries, a lithium-ion battery is constructed of alternating plates, or electrodes of dissimilar materials immersed in an electrolyte. In the typical lithium-ion battery, the anodes (negative electrode) are often made from graphite doped with carbon. The cathode plates (positive electrode) can be manufactured a number of different ways, including from layered oxides (e.g., cobalt), from polyanions (e.g., iron phosphate), or from spinels (e.g., manganese oxide). These positive and negative electrode plates are separated by a material such as plastic or propylene that blocks the flow of electrons but is permeable to the movement of ions. An electrolyte medium (typically a lithium salt dissolved in an organic solvent) is used to immerse the plates in. Positively charged lithium ions are created and move through the electrolyte and separator from the anode to the cathode. The creation and movement of the positively charged ions means that free electrons are also created. These electrons gather on the anode, where they can be conducted away and used to flow through an electrical circuit, doing useful work.

There are a number of different types or variations of lithium-ion batteries available. These types are frequently categorized by the material that is used in their cathode. The most common consumer grade of lithium-ion batteries are lithium cobalt oxide ($LiCoO_2$) and lithium manganese oxide ($LiMn_2O_4$), which offer some of the highest specific energy densities available in consumer batteries. This is the primary reason they are used in products such as laptop computers. The downside of these particular types of lithium-ion batteries (besides their high cost) is their sensitivity to temperature swings, over-and undercharging, and the danger of exploding or burning when subjected to these conditions. For this reason alone, these specific types of lithium-ion batteries are not recommended for use in electric vehicles.

$LiFePO_4$ Battery—Fortunately, there is an excellent lithium-ion battery that has all the positive attributes of other lithium ion batteries but none of the safety concerns: the lithium iron phosphate, or $LiFePO_4$ battery. In a typical $LiFePO_4$ battery, the anodes (negative electrode) can made from copper foil or plate that is doped with porous carbon. The cathode plates (positive electrode) are made from aluminum that are coated with lithium and often doped with other materials. These positive and negative electrode plates are separated by propylene, and a lithium salt-based organic solvent electrolyte is used. While $LiFePO_4$ batteries are not quite as energy dense as other lithium-ion batteries, they are still significantly better than lead-acids; they pack three to four times as much energy (e.g., 100–160 Wh/kg vs. 30–40 Wh/kg). They are also capable of being cycled (charged and discharged) many more times than a lead-acid battery before requiring replacement, and their self-discharge rate (i.e., the measure of how they perform after being left unused for a long

A small 48-volt, 20 amp-hour $LiFePO_4$ battery in an electric bicycle. Often referred to as "duct tape" batteries (because of the way they are constructed), $LiFePO_4$ batteries are an affordable choice for applications like bicycles, scooters, and motorcycles.

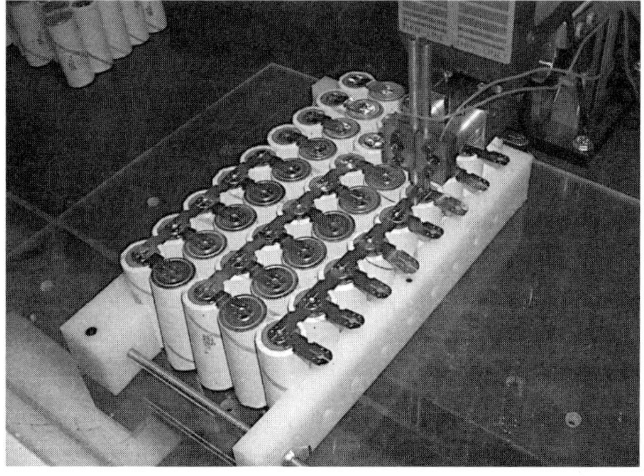

Most OEM-type electric vehicles, such as those made by Toyota and Honda, utilize a large number of small lithium batteries wired together to form a large battery bank. For the average DIY electric vehicle constructor, this approach is not practical. For some specialty vehicles, however, this approach can yield spectacular results. Shown here are a group of A123 lithium nanophosphate batteries being assembled for use in a high-speed drag race motorcycle, which routinely runs a standing start quarter mile in under eight seconds flat. Courtesy KillaCycle.com.

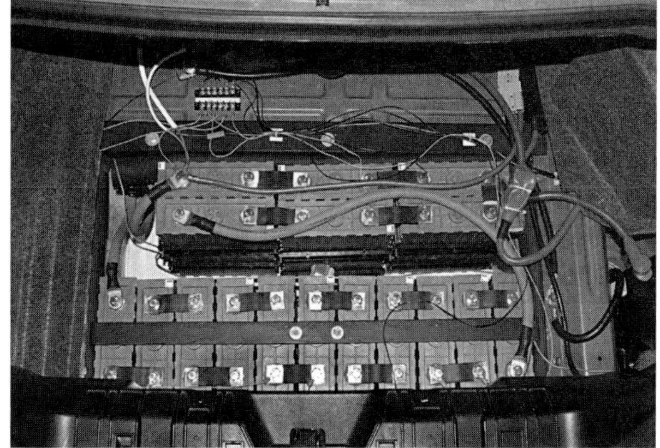

A set of Thundersky lithium-ion batteries installed in the rear trunk of an electric BMW roadster. Note the copper straps that are used to connect one battery to the next. Courtesy Tim Catellier.

period of time) are considerably better. They also can be discharged much more deeply than a lead-acid battery without affecting life or performance.

All these benefits, however, come with a price. Lithium iron phosphate batteries are relatively expensive. They also require expensive battery management systems to charge and equalize the individual cells. This means a high initial outlay of cash is required to fit an EV with $LiFePO_4$

A set of 40 amp-hour LiFePo4 battery cells mounted into an AWD race vehicle. Total system voltage of the 110 battery cells is roughly 360V. Courtesy Dennis Palatov and DPCars.net.

batteries. The good news, however, is that the long cycle and calendar lifetime of the battery, when compared to the shorter lifespan of a lead-acid battery, makes the $LiFePO_4$ a viable long-term choice.

There are a number of high-quality manufacturers of $LiFePO_4$-type batteries. One of the most popular is the Thundersky brand, which makes large prismatic batteries proven to work extremely well in high-performance electric vehicles. Other manufactures, such as A123, also make very high-quality $LiFePO_4$ batteries. Unfortunately, these latter units are typically available only in small D-size units, which are more suitable to applications like cordless drills than large EVs. Some large-scale manufacturers (such as Honda, Toyota, and Ford) are developing systems that utilize hundreds, if not thousands, of these smaller A123-type cells. For one-off home-built applications, however, this type of solution is not recommended. The sheer number of connections that need to be implemented perfectly makes this system impractical for most home builders. For a large-scale manufacturer like Toyota, with production line capabilities and high-quality control abilities, making and managing all the connections is not an issue; for do-it-yourselfers, the problem is not so easily tackled. Make even one poor connection and the entire system may not function properly. High resistance, corrosion, and vibration are the enemies of battery connections. For this reason alone, it's strongly recommended that large Thundersky-type batteries be used in DIY-type EV projects.

Other Battery Types

There are dozens, if not hundreds of other chemistries, configurations, and battery types available for use in secondary (i.e., rechargeable) applications. For example, nickel-metal hydride (NiMH) batteries have been available for decades and might seem like an attractive choice. Further, they are the most widely used battery type in commercial hybrid BEV/ICE vehicles, such as the Honda Insight and the Toyota Prius. This might seem to make them an attractive choice for an EV conversion. Wouldn't it be cheap and easy to find a wrecked hybrid at a junkyard and use its batteries in your own conversion project?

Unfortunately, there are some good reasons not to consider this approach. For one thing, the sheer number of batteries used in these applications means that the battery bank requires a sophisticated charging and management systems. Also, most of these types of applications utilize hundreds, if not thousands of small NiMH batteries. Each has to be individually wired up and checked that is has low resistivity and no chance for corrosion to set in. It's also difficult to evaluate every single battery for capacity and remaining life. One bad battery can affect the life and capacity of the entire bank. Finally, without knowing the history of the vehicle, it's impossible to ascertain how hard the system was abused by the previous owner. What may seem like a great deal might in fact be a waste of money. For these reasons alone, these types of batteries are not recommended for your first EV conversion project.

There are a number of other chemistries and configurations that also might be tempting to use in an EV conversion. These include nickel-cadmium, nickel-zinc, silver-zinc, zinc-air, and so on. Some enterprising backyard EV constructors have experimented with some or all of these systems. There undoubtedly have been some good results, but the vast majority of these experiments ended up with poor outcomes. The problems usually included poor battery memory effects, short life spans, high self-discharge rates, the need for complicated battery charging and management systems, and/or high sensitivity to temperature swings. If we also factor in concerns such as the environmental hazards associated with substances like cadmium, it's clear why most experienced EV builders shy away from alternate battery types. For the vast majority of EV builders, lead-acid or lithium-ion batteries offer the most reliable, robust, and affordable solution to powering their EVs.

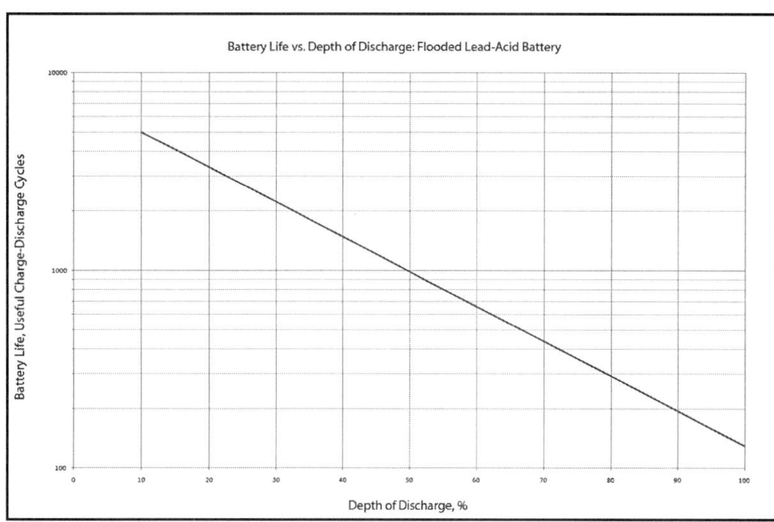

A major factor that affects battery life is how deeply the battery is discharged prior to being recharged. Shown here is a logarithmic plot of useful battery charge-recharge cycles versus depth of discharge (DOD) of the battery. Typically, the most economical usage of lead-acid batteries in EVs is at an average DOD of around 50%. Every battery has a finite life, and every discharge-recharge cycle uses up some of this life. The deeper the discharge, the more of that life gets used up. An EV with few batteries installed means higher DODs and shorter overall life spans. In contrast, an EV with many batteries means lower DODs (and hence higher cycle lifetimes) but also means that the initial battery bank purchase cost is much higher, and the EV will be forced to carry significantly more weight, which in turn requires more energy to accelerate. Both theory and practical experience show that sizing a battery bank so that it will be discharged to 50% DOD during normal commuting is the most economical solution.

Battery Capacities, Ratings, and Life Cycles

Most of us know about car battery ratings from advertisements on television or radio. Cold cranking amps, reserve capacity, and cold/hot weather starting power are some of the marketing terms bandied about by fast-talking advertisers. This battery has X cold cranking amps, while that battery has Y reserve capacity. What do these numbers mean, and, more important, do they even matter when it comes to selecting batteries for an EV project?

CCAs—Cold cranking amps, or CCAs, are a standard measure of car battery performance. It is defined as the number of amperes a battery can produce for thirty seconds when the temperature is 32 degrees Fahrenheit. While important for a starting, lighting, and ignition (SLI) battery used under the hood of an ICE-powered car, CCAs are essentially meaningless for evaluating EV storage batteries. Now let's look at some terms that actually matter when it comes to evaluating EV-type batteries.

Capacity—In simple terms, a battery's capacity is a measure of the amount of energy it is capable of internally storing and releasing while staying above some nominal output voltage (called the *end point* voltage).

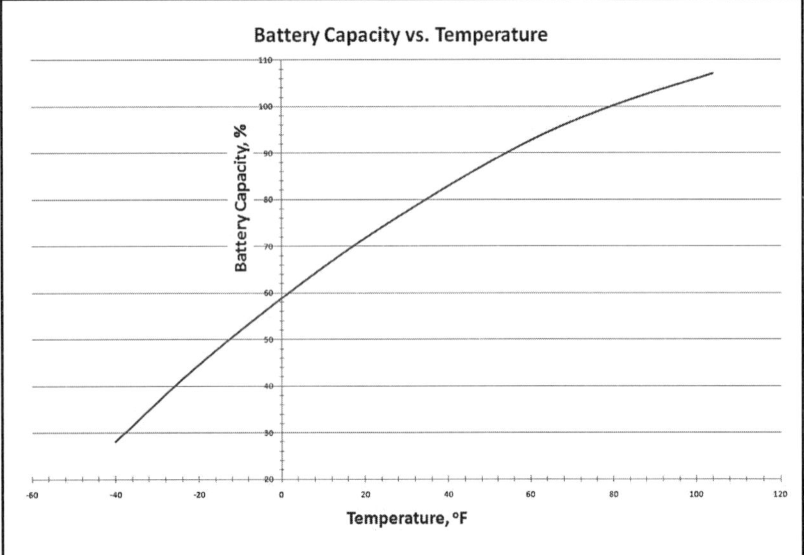

Battery capacity depends on the ambient operating temperature. The colder the temperature, the lower the capacity. Higher operating temperatures, while providing for larger battery capacities, also result in lower overall battery life spans. There is no free energy lunch for EV owners in hot climates.

Discharge Current (amps)	C-Rate	Discharge Time (hours)	Voltage At End of Discharge
0.5	0.05C	20	1.75V/cell
0.1	0.1C	10	1.75V/cell
2	0.2C	5	1.70V/cell
2.8	0.28C	3	1.64V/cell
6	0.6C	1	1.55V/cell
10	1C	0.5	1.40V/cell

If a battery were a perfect energy delivery device and behaved linearly, a 10 amp discharge would take, say, 2 hours to discharge. At a 20 amp discharge, the same battery would take 1 hour to discharge. Batteries are not perfect however, and the rate at which they are discharged affects how much energy the battery can provide. The discharge characteristics of a battery are often expressed in terms of its C-rate.

The *available capacity* of a battery depends in part on how fast you attempt to discharge it relative to its *total capacity*. Available capacity is always less than total capacity.

The capacity of a deep-cycle battery is often stated in units of ampere-hours. One amp-hour is one amp flowing for one hour, or 10 amps for 1/10 of an hour and so on. In other words, an amp-hour is amps multiplied by hours. If you have a circuit (e.g. a motor) that pulls 20 amps from a battery, and you use it for 30 minutes, then the amp-hours used would be 20 (amps) x 0.5 (hours) = 10 amp-hours, or 10 Ah.

Most deep-cycle batteries are specified on a "20-hour rate" basis. This means that the battery is discharged from its state of maximum charge down to its end point voltage over a 20-hour period while the total actual amps it supplies are measured. The number of amps supplied during this period of time are multiplied by the 20 hours, and the result is the Ah rating of the battery. Sometimes ratings at the 5-, 6-, 8-, 10-, or even 100-hour periods are also given for comparison and for different applications. The 5- and 8-hour rates are often used for industrial batteries because these are the typical daily duty cycles of such a unit.

If a battery were a perfect energy source it would behave linearly, independent of load or discharge rate. Batteries aren't perfect energy sources, however. The faster energy is drawn from a battery, the faster it loses its ability to maintain voltage, and the effect isn't linear. This non-linear effect is known as the Peukert effect.

Electric vehicles normally require energy to be drawn from a battery on a fast basis, but most manufacturers don't rate their batteries on less than a 5-hour basis. Some manufacturers, in fact, only rate their battery capacities on a 100-hour basis, which makes the battery appear to have a higher capacity than it actually will in practice. Unfortunately, converting a battery rating from something like a 100-hour basis to a much shorter time scale basis is difficult at best (due to the aforementioned non-linear effect). As a result, some educated guesswork is often required to estimate how many amps can be delivered when the time basis is significantly lowered. For example, a Trojan 8-volt deep-cycle T-875 battery is rated at 170 Ah at a 20-hour discharge but is rated at only 145 Ah if the discharge time is 5 hours. But if it's expected that the battery will need to be fully discharged in, say, 2 hours, the result will not necessarily be proportionally less. If in doubt, talk to the manufacturer of the battery or with other EVers who have used similar batteries in the past.

Another thing to note about battery capacity is that it is often reduced as temperature goes down.

Ampere-Hour

An ampere-hour is the total amount of electrical charge transferred when a current of one ampere flows for one hour. Therefore, the total usable charge stored in a battery can be stated in terms of ampere-hours. In other words, ampere-hours specifies how long a current of a particular amperage can be supplied by the battery.

> ### End Point Voltage
> The end point voltage of a battery is a minimum voltage level used to determine when the battery is discharged. For a typical lead-acid battery, an end point voltage threshold of roughly 1.65 to 1.75 volts per cell is fairly standard. Batteries can be discharged beyond this point, but damage may occur and/or the battery may be unable to be fully recharged and/or hold a full charge again. Life may also be compromised.

This is why an SLI-type car battery often dies on a cold winter morning, even though it may have worked fine the previous evening. If your EV will be used in cold weather, this reduced capacity effect needs to be taken into account when sizing the system. Most batteries are rated at room temperature (approximately 77 degrees Fahrenheit). At freezing, capacity can be reduced by 20% or more. Conversely, capacity tends to increase with higher temperatures. Unfortunately, there is no free lunch for someone living in a warm climate; higher temperatures increase capacity, but they also decrease overall battery life at the same time.

State of Charge—The amount of energy currently held within a battery is known as its *state of charge*, or SOC. In layman's terms, the SOC of a battery is the equivalent of a fuel gauge reading of the battery at a specific point in time. State of charge is normally expressed in percentages; a completely full and charged battery is said to be at 100% SOC, while a fully depleted battery is said to be at 0% SOC. State of charge can be determined a number of different ways. The most common methods are a) voltage; b) current integration; and c) chemical.

The voltage method converts a measurement of the battery voltage to SOC by way of a known discharge curve. Unfortunately, battery voltage is significantly affected by a number of external factors, such as temperature, as well as the current draw from the battery. There are a number of methods to account for these factors (e.g., by way of so-called look-up tables), but for most do-it-yourself applications, voltage measurements provides just an estimate of state of charge.

The current integration method (which is also known as *coulomb counting*) calculates the battery state of charge by way of continual measurement of the current flow to/from the battery and integrating it over time. If a meter is calibrated properly, this technique can give a fairly accurate picture of SOC.

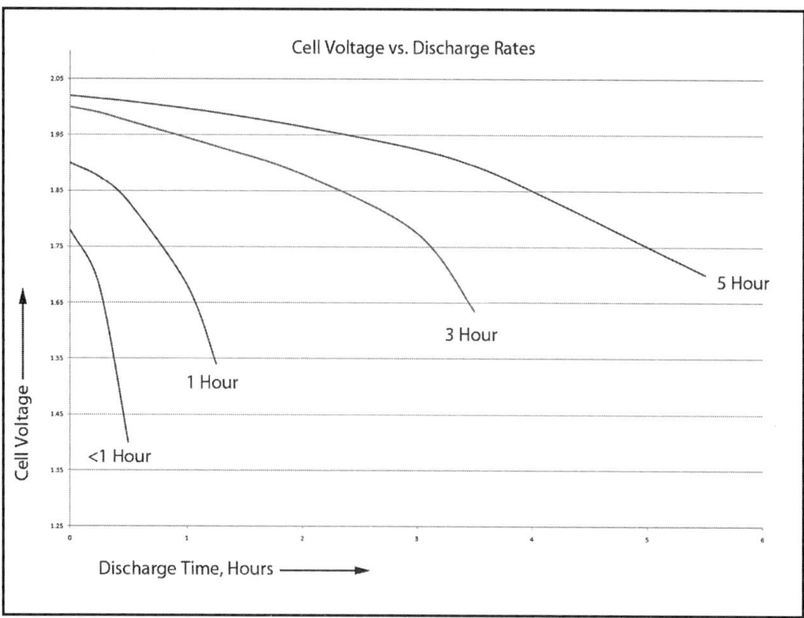

Batteries can either be discharged slowly or quickly. If you discharge a battery slowly (i.e., draw a low amount of current out of the battery over a long period of time), the battery can supply more overall energy than if it were discharged quickly (i.e., draw a high amount of current over a relatively short period of time).

The chemical method of determining a battery's SOC involves measuring the pH of the electrolyte. As a lead-acid battery discharges, the concentration of the sulfuric acid electrolyte is reduced. This in turn reduces the specific gravity of the solution in direct proportion to the SOC. A hydrometer can be used for this purpose and, if performed carefully, can produce a fairly accurate determination of state of charge. The trick to this, of course, is that the battery needs to be a flooded type, and the individual cells need to be fitted with caps that allow access to the fluid.

Depth of Discharge—A compliment to a battery's state of charge is something called the *depth of discharge* (DOD). The DOD of a battery is a measure of how much energy has been removed from a fully charged battery, which can be measured in either ampere-hours, or as a percentage. As the SOC drops, the numerical value of the DOD increases proportionally. A battery that has an 80% state of charge, for instance, would be considered in a 20% depth of discharge condition. Both SOC and DOD are often used to express the energy content of a battery.

Battery Cycles—When a fully charged battery is discharged and then recharged back to full capacity, it is said to have experienced one complete battery cycle. Some batteries, such as lithium iron phosphate types, can be discharged down to nearly zero capacity. Others, like normal flooded cell lead-

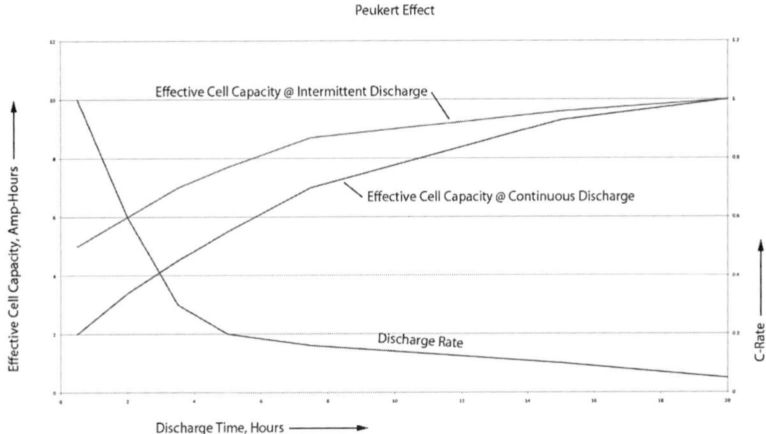

Battery efficiency is often expressed by a *Peukert* number, which represents the internal resistance of the battery. Low numbers indicate efficient batteries while high numbers indicate high internal resistance and low efficiency. A number close to 1 indicates a well-performing battery. This chart shows how effective cell capacity varies with discharge rate. Note that an intermittent discharge improves overall capacity. This is because the chemical reaction in the cell has a nominal amount of time to recover in between load application.

acid batteries, should be discharged less, which helps improve lifespan; typical depth of discharge for a lead-acid battery used in an EV is 80%. This means that a full cycle would be 100% SOC discharged down to 20% SOC, followed by recharging back to 100% SOC.

Battery life is directly correlated to how deeply the battery is cycled. If a battery is discharged to 50% every time it is used, it will last about twice as long as if it is cycled to 80% DOD. If the battery is cycled only to 10% DOD, it will last roughly five times as long as one cycled to 50%. The obvious trade-off then becomes one of increasing the number and size of batteries in an EV while minimizing overall weight and cost. There are clear practical limitations for this; it's silly to install twice as many batteries than normal in an EV just to cut the DOD by 50% and, therefore, increase the battery bank life span by a factor of two.

When evaluating batteries for use in an EV, it is important to fully understand what DOD was assumed when the battery was rated by the manufacturer. There are occasional reports of unscrupulous vendors that advertise batteries with very high cycle abilities but neglect to point out that the battery was only discharged 10–20% during the testing.

Selecting Batteries for Your EV

Step 1: System Voltage—Before selecting a type and individual size of battery for an EV, you need to determine the overall system voltage. While factors such as gearing and aerodynamics certainly come into play, the primary electrical limitation on top speed of a conversion is the total voltage of the battery pack. More volts equal higher top speed; the more batteries you wire together in series, the higher the speed you can achieve. Electric motors create something called *back-EMF*, which is essentially a reverse voltage that the battery bank has to overcome in order to drive the motor. The faster the motor turns (i.e., the faster the car travels), the higher this back-EMF effect. In a sense, voltage determines the maximum rpm possible from the motor, provided there is sufficient current to overcome the load that the motor is driving.

While it is possible to calculate the required voltage of a specific application, the analysis is beyond the scope of this book. Power requirements of a car increase as a function of the speed cubed. Also, factors such as rolling resistance, aerodynamic drag, drivetrain losses, and specific motor back-EMF operating characteristics all have to be calculated and added into the analysis. Without sophisticated computer modeling software, wind-tunnel test results, and/or access to advanced electric motor EMF data, precisely calculating required voltage comes down to educated guesswork.

Okay, that's the bad news. The good news is that you can do a pretty good job estimating system voltage requirements by simply looking at what others have done before you on similar vehicles conversion projects. A great place to start is by perusing the project vehicles on evalbum.com, where system voltage and measured top speeds are usually included in the vehicle descriptions. There are also several good rules of thumb. For instance, in order to travel at 50 mph in a small, relatively aerodynamic car, you probably need at least 96 volts. If you want to travel at 60 mph, however, you probably need at least 120 volts. The more un-aerodynamic your vehicle is (e.g., a pickup truck), the more voltage will be required. A small Chevy S-10 pickup, for instance, will probably need at least 144 volts to achieve 60 mph. At the other end of the scale are small vehicles like motorcycles and bicycle conversions. A motorcycle conversion, for instance, will usually be able to achieve 60 mph with 60 volts.

Step 2: Battery Type—Once an overall system voltage has been selected, the next step in determining the type of battery for your EV project is to reevaluate how much money you have budgeted for this part of the conversion. It sounds simplistic, but it's true nonetheless: cheap batteries typically don't have as much capacity as more expensive batteries. At the low end of the scale are the flooded lead-acid *traction* batteries. Next up the line are the sealed and AGM-type lead-acids. At the high end are lithium-based batteries. If money is not a big consideration, lithium iron phosphate batteries are probably the best choice today. They are more compact, lighter in weight, and have higher specific energy densities than comparable lead-acids. They also can be mounted in places and orientations that owners of flooded-type equipped EVs can only dream of. The downside, beside costs, is the need for a battery management system. Lithium-type batteries are also somewhat more difficult to find on the market than flooded lead-acids, which can be found in almost any town. Reputable manufacturers include Thunder Sky and Ping, which are both Chinese companies.

If money is a more serious consideration, flooded lead-acids are probably your best choice for both the first-time EV project, as well as standard commuter-type vehicles. These types of batteries are cheap and durable, highly recyclable, available in a wide variety of sizes and shapes, and able to withstand some over- and undercharging without suffering serious damage. Many new EV owners ruin a battery or two at first as they figure out the ins and outs of maintenance and charging, but the relatively low cost these batteries compensates for this issue. The main problems with these types of batteries are relative bulk, weight, and accessibility requirements. Flooded lead-acid batteries require periodic testing and watering to maintain proper performance and maximize life. There are a number of reputable brands of flooded lead-acid batteries. Two of the most common are Trojan and U.S. Battery, both of which make high-quality units in a variety of sizes, voltages, and shapes.

Absorbed glass mat (AGM) and gel-type sealed deep-cycle lead-acids are a good choice for more experienced builders and/or for higher-performance applications. These types of batteries can deliver relatively high currents with lower voltage sags than flooded-types. They are also somewhat smaller and more compact than equivalent flooded batteries, are essentially maintenance-free, and can be mounted in any orientation, even upside down. They do cost more than flooded batteries but are still considerably cheaper than lithium-type batteries. AGMs are also somewhat sensitive to overcharging; to extend their lifetime, AGM battery packs should ideally incorporate some type of battery regulators or balancers, such as those made by Manzanita or AIR. Many moderate cost, high-performance EVs have been successfully built using AGM batteries. Reputable manufacturers include Deka, Hawker, Optima, and SVR.

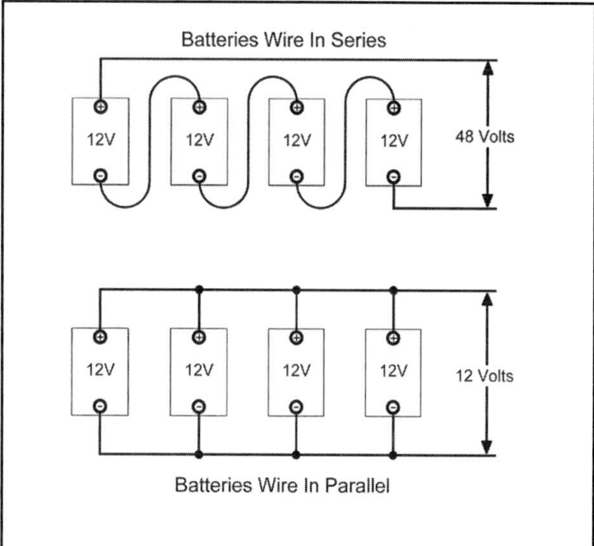

Batteries can be connected together in various ways to increase voltage or current output. In the top circuit, four 12-volt batteries are connected in series, which results in an overall system voltage of 4 x 12 volts = 48 volts. Current flow in this series arrangement does not increase, however; the current flow from one battery is equal to the amperage from the entire four batteries. In contrast, the bottom circuit is wired in parallel, meaning that the overall system voltage is still 12 volts, but the current flow ability is four times higher than a single battery.

Step 3: Battery Sizing—The next step is to take a look at the range you need. This will then help you determine the specifications of the battery type that best suits these needs. It will also help you decide how many of these batteries will be required to achieve this range.

As we saw earlier, batteries are often rated in terms of their capacity, which is expressed in ampere-hours. If we multiply a battery's nominal voltage by its ampere-hour rating we get a measure of the battery's total available energy, expressed in units of watt-hours. This value is the theoretical amount of energy that we can withdraw from a single battery. If we then multiply this amount by the total number of batteries in our system, we have the total theoretical amount of energy stored. Unfortunately, there are a few things that reduce the actual amount of energy we can withdraw from the bank. For instance, most battery manufacturers recommend that lead-acid batteries be discharged to a maximum of 20% of their capacity. Deeper discharges result in overall fewer charge-discharge cycles possible (i.e., battery life). Also, as we saw earlier, the rate at which a battery is discharged can greatly affect the output of the battery. For these reasons, a conservative rule of thumb is to assume that only 50–60% of the theoretical available

Mixing Old and New

As a general rule, you should not mix old and new batteries together in an EV conversion. You should also try to avoid using mismatched or different types of batteries together. Different batteries charge at different rates and draw/supply different levels of current. If you charge a mixed pack with a simple series charger, some batteries (especially older ones) will tend to be overcharged, and others (any newer batteries) will be undercharged. The result is a reduction in the overall battery pack lifespan. Separate chargers and/or expensive battery management systems are required if you're using different types of batteries together.

Different types of batteries have different capacities, too, including different discharge rates and the ability to hold a charge. This is also true of old vs. new batteries of the same type. What this means is that the overall range of a vehicle is likely to be significantly reduced, as the system capacity is dominated by the weakest battery in the pack. In a sense, adding a smaller or weaker battery to a system of larger/stronger batteries simply brings the larger batteries down to the level of the weak link. This means they act as nothing more than dead weight and space in the car.

If money is an issue when buying parts for your EV, it's recommended that you hold off purchasing the batteries until the end of the project, and then only after you've saved up enough to buy all the batteries together in one shipment. If necessary, purchase the batteries on a payment plan, where you don't take delivery until it's entirely paid off, thereby receiving a full set of equally fresh batteries ready to be installed at once.

The only time you should mix batteries is if one or two of your batteries are damaged or are defective and you have to replace them. If this happens to you, be sure to replace the bad units with batteries that have similar age and usage history if you can find them. Remember, one bad apple (battery) will hurt the entire bunch (battery bank).

energy content of a lead-acid battery bank is actually available for use in propelling an EV. If you are using lithium-based batteries, you can assume that more energy can be withdrawn from the battery (approximately 75–80% or more).

So now that we know how much energy is available in a battery bank, the next step in calculating how far we can travel is to determine how much energy is required per mile of usage. The bad news is that to precisely calculate this Wh/mile figure we have to know a lot of information and make detailed engineering calculations. For example, the amount of energy required to move a vehicle a certain distance is dependent upon the terrain (i.e., hilly vs. flat), the weight and aerodynamics of the vehicle (i.e., whether the car is easy or hard to accelerate and maintain speed), and how heavy the right foot of the driver is.

Okay, now for the good news. There are approximate rules of thumb and empirically derived values that that can be used to estimate how far typical EVs can go on a charge. For instance, "normal" cars or light trucks that have been converted to electric power typically consume between 200 and 400 watt-hours per mile (Wh/mile) of "normal" usage. Motorcycles use less than this amount, often in the range 75-150 watt-hours per mile. Of course if you drive up long hills at high rates of speed in a heavy EV, your Wh/mile consumption rate may skyrocket to 2, 5, or even 10 times these levels. (Conversely, electric bicycles can be as low as 20–30 Wh/mile, or even lower if a lot of pedaling is used in conjunction with the power assist!)

So let's now look at a simple example. Let's say that we're building a small and efficient commuter car like a Geo Metro. The batteries we selected are 12V flooded-type deep-cycle lead-acid batteries. The battery manufacturer states that the 20-hour capacity of the battery is 150 amp-hours. We will be using 10 of these batteries in our conversion.

Now, we know from our previous discussion that the actual amp-hour capacity of the battery will be less than the 20 hour figure when used in a high-draw application like an EV. Let's assume something like 120 Ah. Our total energy storage ability of our battery bank is: 10 batteries x 12 volts x 120 Ah =14,400 Wh. To account for allowable depth of discharge and other factors, we will divide this figure in half to determine a conservative figure for the amount of usable stored energy in the battery bank: 14,400 ÷ 2 = 7,200 Wh.

The next step is to determine how many watt-hours are required to move this vehicle one mile. To help determine this, we peruse similar cars at EVAlbum.com and see that Geo Metros, driven on flat and level ground, average around 250 Wh/mile. We can then simply take the 7,200 Wh value and divide it by the 250 Wh/mile figure. This gives us 7,200 ÷ 250 = 28.8 miles of usable range.

Step 4: Battery Size & Packaging—If you've gotten this far, you now have a pretty good idea of the overall system voltage required and the basic battery type and capacity that would meet the needs of your conversion project. At this point, it's useful to consider other factors that may modify your battery choice. Specifically, you need to consider how many batteries will be needed to achieve the required system voltage, and then how you will package all those batteries in the vehicle.

Further, not only does there have to be sufficient room in the vehicle for all the batteries, but there also has to be a means of safely supporting and restraining them from movement. Wiring considerations are also important; how will you connect the terminals together, and how will the wiring be routed? Also, you need to ensure that each battery can be accessed easily for periodic maintenance and testing. This is especially true if the flooded lead-acids are used. It's one thing to be able to reach a battery and pull a cap off for testing the pH of the electrolyte; it's an entirely different problem when you have to get a funnel into the open cell aperture and add water.

Weight is another consideration. You will want to ensure that the overall vehicle GVWR is not exceeded, and that the overall front-to-rear balance of the car is not grossly different than the stock vehicle.

Many constructors of electric vehicles find that building simple cardboard mockups of batteries is a useful exercise. Batteries are heavy. It's much easier to build four or five simple fake batteries and place them in various configurations under the hood to determine the best layout of the overall system.

When planning an EV, care should be taken to ensure the units are properly supported and held in place. Shown here is a typical angle-iron frame used to hold a set of flooded lead-acid batteries. Frames like this are available for purchase for some popular vehicle models, such as the Chevrolet S10 and Porsche 914. If a supplier doesn't make an off-the-shelf battery rack for your car, you will have to construct your own. Courtesy Electric Car Company of Utah.

A set of B.B. Battery HR22-12 VRLA-type maintenance-free lead-acid batteries mounted in an electric motorcycle conversion project. Total system voltage is 72 volts, with each battery able to supply 22 amp-hours. Note the quality welds and well-engineered mounting details. Batteries are heavy, so mounting them low and compactly like this in the center of the chassis helps improve handling. Spending the extra time to build a solid and accessible mounting system like this is extremely important for both safety and performance. Courtesy Tony Helmholdt.

Small pickup trucks are popular for conversion in part because the batteries can simply be mounted in the bed. The downside is that usable hauling space is taken up. Care must also be taken to securely hold the batteries in place via hold down clamps. Courtesy Electric Car Company of Utah.

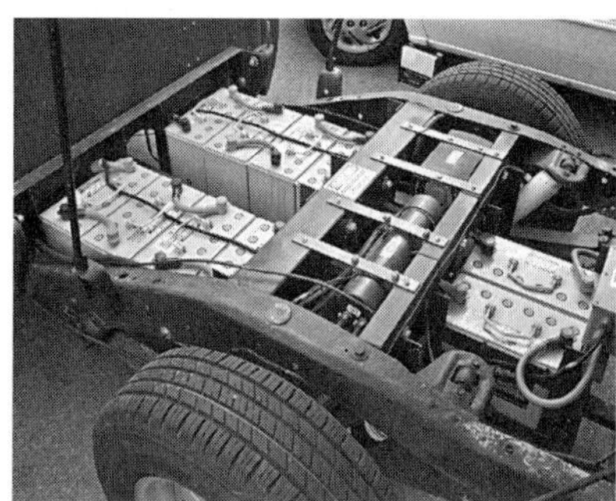

Many builders of electric pickup trucks mount their batteries underneath the bed, inside the frame rails. Some go so far as to modify the bed so that it swings up and out of the way for servicing the batteries. Courtesy Tony Helmholdt.

Batteries are heavy, and it can be difficult to move them around and temporarily support them in place underneath the hood of a car or truck when working out the mounting details. A solution that many builders use is to create lightweight cardboard or plywood battery mock-ups that are the same size and shape as the actual battery. This allows test fitting to take place without throwing out your back or causing a hernia. Courtesy Tim Catellier.

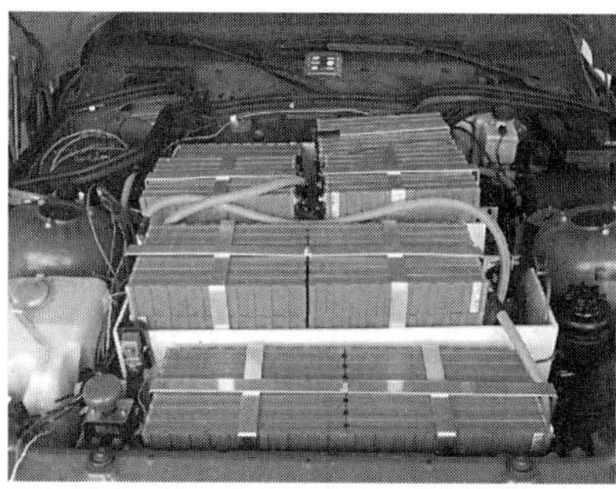

Here is the same vehicle with its final under-the-hood mounting configuration. The batteries in this application are Sky Energy 120AH, 3.30 Volt Lithium-Ion units. Courtesy Tim Catellier.

Chapter 7

Battery Chargers

"Batteries are like people. They need to be recharged on a regular basis to stay healthy." —Anonymous

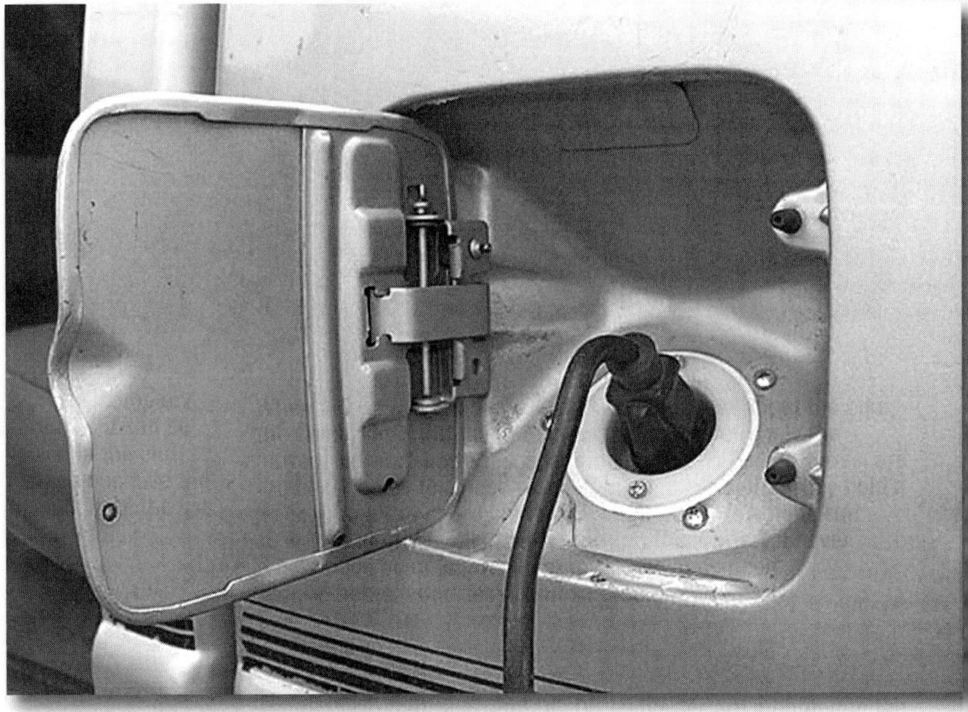

Many inexperienced EV owners ruin their first set of batteries within the first few months of commissioning because they didn't understand the basic do's and don'ts of battery recharging. Batteries are expensive, and learning how to recharge by trial and error can be a costly undertaking. Courtesy Electric Car Company of Utah.

The topic of battery charging may not be particularly exciting for most new EV owners, but it's one of the most important subjects to understand. Safely and efficiently transferring electrical energy into the batteries after use is vitally important to the performance—and reliability—of the vehicle. Perhaps even more important, the means by which a set of batteries is charged can have a significant effect on its lifespan. Charge a set of batteries incorrectly, and the number of useful battery charge-recharge cycles possible can be radically shortened. In fact, many new, inexperienced EV owners ruin their first set of batteries within the first few months of commissioning because they didn't understand the basic do's and don'ts of charging. Batteries are expensive, and learning how to recharge by trial and error can be a costly undertaking. Often, the reason you see used EVs for sale with advertisements that say "needs batteries" is because the owner treated the batteries badly.

The good news is that proper battery recharging is not difficult. The key is in understanding the process by which energy is transferred back into a battery cell and then applying the correct technology and methods to ensure this is done in a safe and reliable manner. (Note that the majority of this chapter focuses on flooded lead-acid battery charging as this is by far the most common type of battery used in EV conversion projects. Other batteries, such as sealed lead-acid, gel-types, and lithium-ion units, are touched on where applicable, but it is recommended that you contact the manufacturer for more information on how to charge these alternate types of batteries.)

Cell Chemistry and Specific Gravity

When a fully charged lead-acid battery is connected to a load, such as a motor, a flow of electricity is created. As we saw earlier, the source of this electrical current is a chemical reaction inside the cells of the battery. In simple terms, the reaction between sulfate (from the sulfuric acid-based electrolyte solution) and the battery's internal lead plates causes the release of electrons, which creates both a potential difference (i.e., the voltage difference across each battery cell) and the source of the electrical current itself (i.e., the electrons). As the battery discharges, lead sulfate and water slowly build up in the acid. Eventually, the electrolyte becomes diluted enough that the supply of electrons slows and finally stops. The battery is said to be discharged at this point.

Fortunately, this chemical discharge reaction in a lead-acid battery is fully reversible. Forcing a direct current back into the battery cell in the reverse direction of the discharge current replaces energy drawn from the battery during discharge. The electrolyte and the lead plates during this charging process act in essentially the reverse manner of the discharge process. Lead sulfate at the plates and the water in

Battery Charging Basics

Basic physics and the laws of thermodynamics show us that the amount of energy required to recharge a battery is more than the energy discharged in the first place. Said another way, the amount of electrical energy it takes to recreate the original specific gravity of a battery cell is more than was released originally by the internal chemical reactions of discharge. If a battery bank produced, say, 100 kilowatt-hours (kWhrs) of useful work during operation, it would take more than 100 kWhrs to recharge it back to its original state. The difference in energy is the amount that is lost in the energy transfer process.

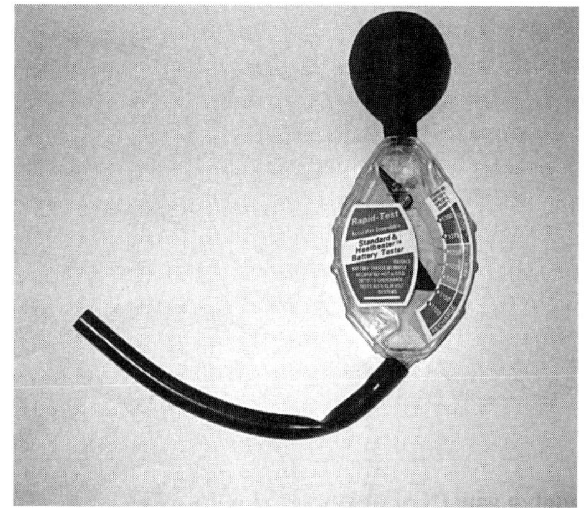

The specific gravity of the electrolyte in a lead-acid battery cell is proportional to the concentration of acid in the electrolyte. If we measure the specific gravity of a battery cell, we can indirectly (but accurately) determine the cells state of charge. The tool most commonly used for this task is a hydrometer.

the electrolyte are broken down into metallic lead, lead dioxide, hydrogen, and sulfate ions. This re-creation of plate materials and sulfuric acid restores the original chemical conditions inside the cell, including a property known as the *specific gravity* of the electrolyte in the battery cells.

Sulfuric acid electrolyte in a lead-acid battery has a certain measurable density, or mass per unit volume. If we divide this electrolyte density by the density of water, we get a measurement called the specific gravity of the electrolyte, or SG for short. The specific gravity is proportional to the concentration of acid in the electrolyte, which is expressed in its pH level. In other words, the specific gravity is a means of determining how potent the electrolyte in each cell is at any given time. If we measure the specific gravity of a battery cell, we can indirectly (but fairly accurately) determine the cells state of charge; as the battery cells discharge, the electrolyte density drops and, therefore, the SG reading will be lower. A fully charged battery cell will have an electrolyte specific gravity of around 1.26 at room temperature. When the cell is fully depleted, the SG will be around 1.10 or so.

A standard 12-volt lead-acid battery is typically constructed with six internal cells. When the battery is fully charged, each of the cells theoretically has a two volt potential, which, when added together, totals up to the 12 volts. In practice, however, each cell has slightly higher than two volts. More important for this discussion, each cell has a slightly different voltage potential than the other cells. Ensuring that each cell is as closely matched to its neighbors as possible is a key to a long, reliable battery life span. Said another way, measuring and monitoring the specific gravity of

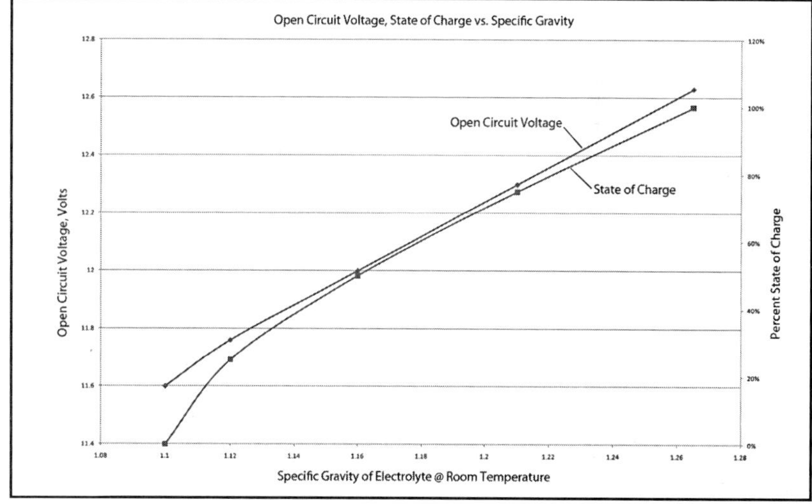

The open circuit voltage (and hence, the state of charge) of a flooded lead-acid battery varies linearly with electrolyte specific gravity, as shown here.

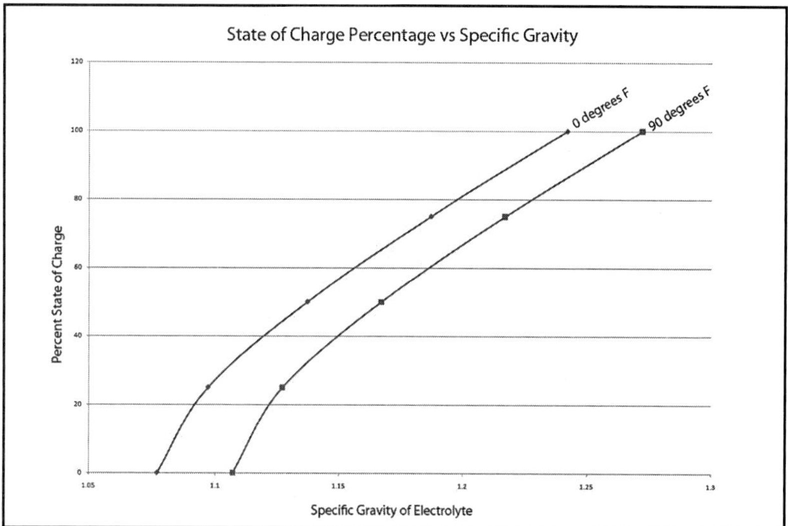

Adjustments must be made when measuring a battery's electrolyte specific gravity to account for the temperature of the fluid. As shown in this graph, there can be a significant difference in specific gravity measurements for two equally charged batteries measured at different temperatures.

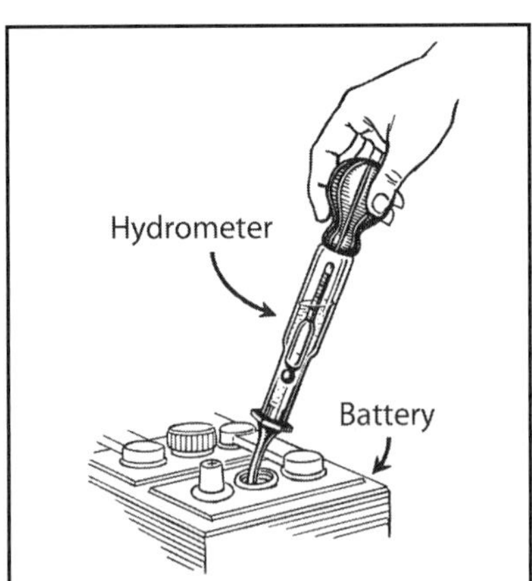

A small amount of electrolyte is drawn out of the battery by the hydrometer. As the battery cells discharge, the electrolyte density drops and, therefore, the SG reading will be lower. A fully charged battery cell will have an electrolyte specific gravity of around 1.26 at room temperature. When the cell is fully depleted, the SG will be around 1.10 or so. Courtesy Pearson Scott Foresman.

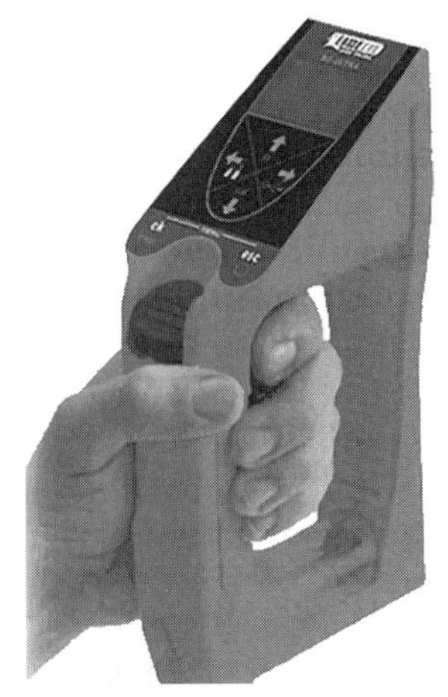

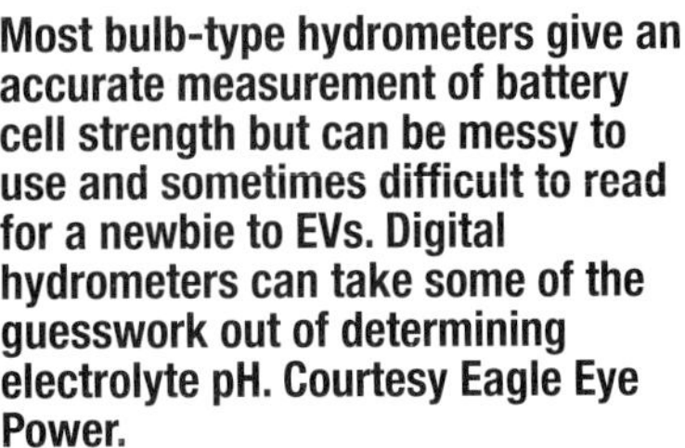

Most bulb-type hydrometers give an accurate measurement of battery cell strength but can be messy to use and sometimes difficult to read for a newbie to EVs. Digital hydrometers can take some of the guesswork out of determining electrolyte pH. Courtesy Eagle Eye Power.

Most garage and automotive consumer-type battery chargers can only provide bulk charging only. These simple units have little, if any, voltage regulation. While fine for providing the occasional boost charge to a starting, lighting, and ignition (SLI) battery used in an ICE-powered vehicle, they are not useful for charging up the battery bank of an EV.

Charging Times

A rule of thumb is that it takes three to five times as long to recharge a lead-acid battery to the same level as it did to discharge it in the first place.

each cell before and after charging will not only tell you the state of charge of each cell but can also go a long way toward ensuring the battery health is optimized during charging.

Three-Step Battery Charging

Proper lead-acid battery charging typically takes place in three distinct steps: *bulk*, *absorption*, and *float*.

Bulk Charging—The first step in a three-stage battery charging process is called *bulk charging*, or *constant current charging*. During this stage, the input current to the batteries is set at the maximum safe rate the batteries will accept until voltage rises to near the fully charged level (80-90% of final charge). This normally takes somewhere between 4-6 hours for a battery discharged to a depth of 80%. There is no particularly correct voltage required for the bulk charging stage, but normally the potential applied to a 12-volt battery is between 10 and 15 volts.

Absorption Charge—Once the battery is charged to 80–90% of its final desired charge, we can begin the second distinct stage of the process, which is called *absorption charging*, or sometimes *constant voltage* or *topping charging*. During this stage, the input voltage to the battery is held constant while the current is gradually tapered off. Voltages at this stage are typically around 14 to 15.5 volts. This stage usually takes between 4–6 hours to achieve, and at the end of the phase the battery may actually be at a 105% or more state of charge. Omitting this step (i.e., simply using the battery at 80–90 state of charge) can eventually cause the battery to lose its ability to accept a full charge. This step also acts as a cell equalizer phase, which helps ensure that all the cells of the battery have nearly identical charges applied.

Float Charge—The final step in the three-stage battery charging is called the *float phase*, or sometimes the *maintenance* or *trickle charge phase*. After a battery reaches its full charge during the absorption stage, the input voltage is lowered to a reduced level, which is typically 12.5-13.5 volts. This step compensates for the batteries tendency to self-discharge, plus allows for the reduction in gassing, as well as helping to prolong battery life.

Chargers, Regulators, and Battery Management Systems

With the combination of modern electronics and low prices, there is no excuse to skimp when shopping for a charger. Selecting (and correctly using) the right charger for your batteries is critical to the long-term enjoyment of your EV. There are a number of factors to consider, but often the first question that needs to be addressed is whether you need an onboard charger or a fixed charging station—or both.

Onboard Charger—An onboard charger is typically mounted under the hood or in the trunk of an EV. It (obviously) travels wherever the vehicle goes, which means charging can take place anywhere there is an available outlet. This type of charging-where-you-park is known as *opportunity charging* and can greatly extend the effective range of an EV. For instance, if your employer allows you to plug into his power system during the day while you work, the result is essentially a doubling of the distance you could travel if you were only able to charge at home overnight. Some enterprising (read: cheap!) EVers have even been known to charge only at their place of business, thereby obtaining all of their electrical needs essentially for free. The chief downside of onboard chargers is that most units are based on 120-volt input, which is fine if recharging time is not a factor but is somewhat less efficient and slower than larger 220-volt systems. Onboard chargers also require physical space under the hood for mounting, and they add some extra weight to the vehicle. For most vehicles, this is not a serious concern, but in high-efficiency, ultra-lightweight vehicles, such as small sports cars and motorcycles, the extra space and weight of an onboard charger can be important.

Fixed Station Chargers—In contrast to onboard chargers, a fixed charging station is a permanent mounted fixture located inside the garage or carport of the EVer's home. The normal operating scenario is that the vehicle is driven during the day, and then returned home to the garage, where it is plugged into the fixed charging station. The batteries are recharged overnight while the owner sleeps. In the morning, the EV is fully charged and ready for a day of commuting. The biggest advantage of fixed charging stations is that they allow for high input voltage, which can improve the overall efficiency of the charging operation. It can also speed up charging significantly. A battery bank that takes 12 hours to charge on 120 volts can be recharged in approximately 50–65% of that time if the charger uses a 240-volt input.

Fixed station chargers do require special wiring in the house, which is an added expense. Ground Fault Circuit Interrupt (GFCI) wiring is normally required, as is compliance with National Electric Code (NEC) and local building code requirements. These types of circuits have both high voltage and high current, so unless you're an expert on electrical wiring it's recommended that you hire a licensed electrician to perform the installation.

A rule of thumb some builders use is that if the battery pack weighs more than 1000 lb, a 240-volt fixed station charger is required. Opportunity charging can still be undertaken with 120-volt input, but a high-voltage system is recommended as the primary means of charging the batteries. Thus the best of both worlds (if you can afford it) is to have both an onboard charger for opportunity charging, and a fixed station charger at home for high efficiency, fast recharging. Once the decision has been made on the type of charger, the next question is what capacity is required in the charger.

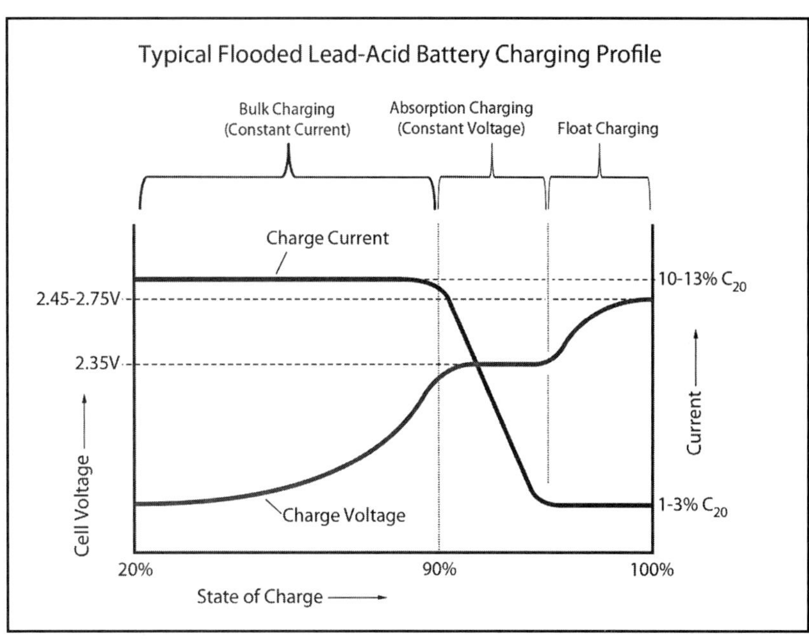

Different types of batteries require different charging schemes. Shown here is the most common method of charging flooded lead-acid batteries. The charge profile takes place in three separate stages, as shown. First is the *bulk* charging phase, where the current is held at a near constant level. Following this is the *absorption* charging phase, where voltage is held constant and the current is slowly ramped down. The third and final step is the *float* charging phase, where charge current is again held constant, but at a lower level than before.

First and foremost, the charger needs to meet the power requirements of the battery make and type. A rule of thumb some builders use is that the charge rate for flooded-type lead-acid batteries should be around 10% of the 20-hour rate. For example, a Trojan 6-volt battery with a 250 Ah rating would need 10% x 250 Ah = 25 amps during the bulk charging phase. If your battery bank is, say, 120 volts (e.g., 20 x 6V batteries), the charger should be able to provide 120 volts x 25 amps = 3000 watts. (AGM and gel batteries require slightly different rates for optimal charging, so check with your battery manufacturer for more information.)

Selecting a Charger—Once the type and size of charger is decided, the next step is deciding the level of sophistication and features you need. At the low end of the scale are simple, manually controlled units, which are often called *bad boy* chargers, or *variacs*. These types of chargers usually consist of a rectifier that converts AC power into DC, which is then either fed directly to the battery or through some type of manual control device. The advantage of this type of system is its low cost. The disadvantages, however, are large, including the need for the operator to fully understand battery charging profiles and then monitor and adjust the charging

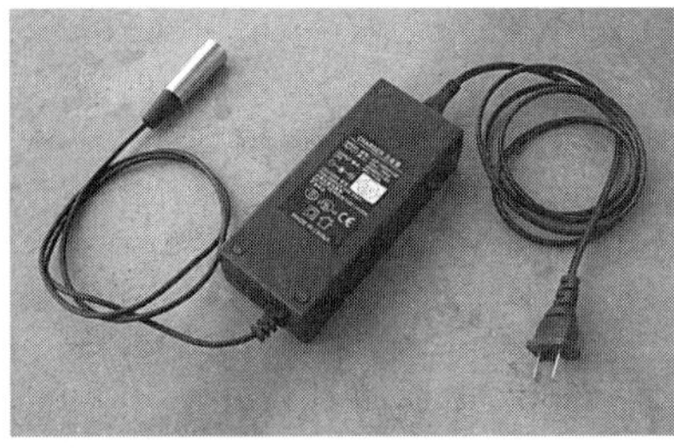

A battery charger must be matched to both the battery type and size to be effective. This small charger is used to replenish a 48-volt 20 amp-hour lithium iron phosphate battery in an electric bike. Total recharge time can take as long as 5 hours if the battery is fully depleted.

Zivan makes a number of high quality battery chargers that are popular with EV owners. Courtesy Electric Car Company of Utah.

Manzanita Micro manufactures a line of popular programmable chargers, such as the PFC-50, shown here, which is capable of providing up to 50 amps of charging output. Courtesy Canadian Electric Vehicles.

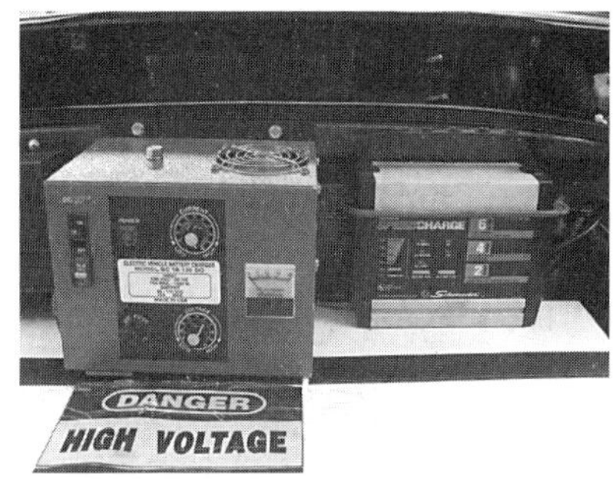

The owner of this electric Porsche 914 sports car uses two chargers. The main charger (shown on the left in this photo) is a Russco SC 18-120 SO 15A 120v charger with digital shut off. It is used to charge the main battery bank of the car. The smaller charger (shown on the right) is a 2-6 amp automatic charger used for replenishing the auxiliary 12V battery. Courtesy David Oberlander.

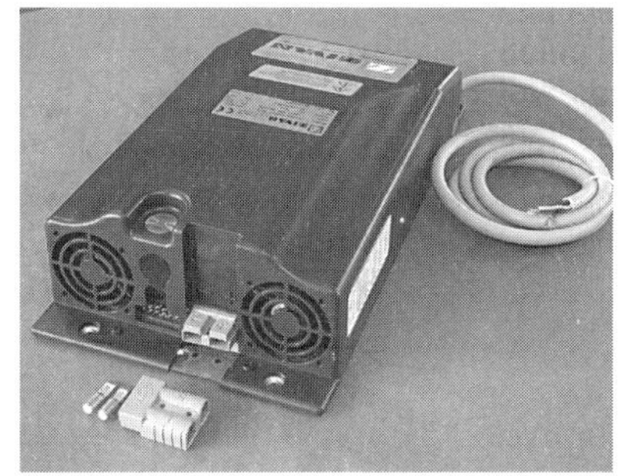

The Zivan NG3 charger is a solid-state transformer-based charger that is popular with EV owners. It can be configured as either a 120- or 240-volt input system. Courtesy KTA Services.

A relatively new battery charger on the market is by NetGain Controls. Note the huge heat sink fins protruding from both side. In addition to a number of features, this unit includes a thermal sensor that can be used for adjusting the charging profile based on actual battery temperature. The charger can be used with both flooded and AGM-type lead-acid batteries. Courtesy EVSource.com.

during the process on a frequent basis. It is strongly suggested that new EV owners stay away from this type of charger for their first project vehicle, as it is very easy to destroy an expensive set of batteries if you don't know precisely what you're doing.

As you move up in charger sophistication the prices tend also to rise accordingly. The secret is determining what you need versus what you want. Fortunately, low-end chargers can work quite well, but you should shop around. Look for a built-in ammeter and some means of manually adjusting the current output. Some chargers have automatic shut-offs, while others need to be manually switched off at the end of charging or rely on timers that need to be set. At the high end of the scale, high-efficiency chargers are available with computer control, preset charging algorithms and/or the ability to be programmed with custom voltage-current profiles. Some come equipped with safety circuits, over-current protection devices, and even temperature probes that can modify the charging profile to compensate for cold or warm weather.

If your battery pack is greater than 120 volts, the charger you select may require the use of a booster transformer or other multiplier device. Others require the use of battery regulators and/or balancers that allow fine adjustment of current and/or voltage to each individual battery in the vehicle, thereby ensuring that each battery is recharged optimally and to its full state of charge, which in turn helps maximize battery life spans. Again, make sure that the charger system you purchase is compatible with the type and size of you batteries.

At the highest end of the battery charger spectrum is something called "battery management systems," or BMS units. These computerized devices monitor individual battery health and operating temperatures, keep them balanced and at the same states of charge, and fully control charging (and even discharging to some extent on some models). A BMS might be considered overkill for a set of lead-acid batteries, but it is often required for more exotic types of batteries, such as lithium-ion and lithium iron phosphate units.

Finally, when shopping for a battery charger you should also look for a good warranty; two to three years is becoming standard with some manufacturers.

Care and Feeding of Lead-Acid Batteries

There are a number of useful dos and don'ts that can go a long way toward ensuring maximum life and performance from a set of batteries. In no particular order of importance, here are some of the

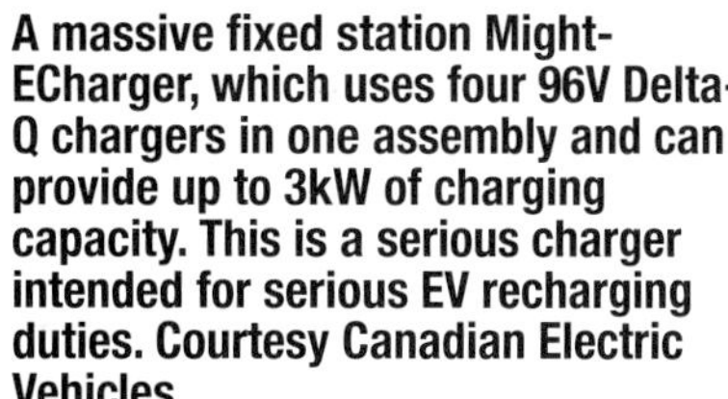
A massive fixed station Might-ECharger, which uses four 96V Delta-Q chargers in one assembly and can provide up to 3kW of charging capacity. This is a serious charger intended for serious EV recharging duties. Courtesy Canadian Electric Vehicles.

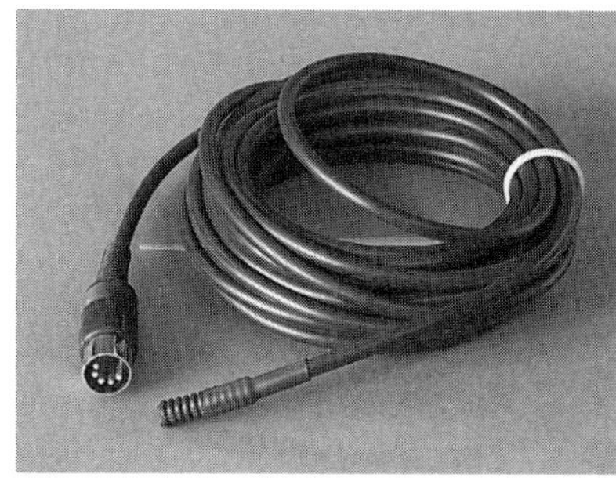
Some battery chargers accept temperature probe inputs to help monitor battery conditions during charging. Shown here is a Zivan probe that interfaces to Zivan chargers. This type of sensor/input system is particularly recommended if you're using sealed batteries. Courtesy KTA Services.

most important things to keep in mind when performing maintenance and/or charging a set of batteries:

Electrolyte—The water level inside each flooded lead-acid battery should be checked on a regular basis. The level should be even in all cells and usually 1/4" to 1/2" below the bottom of the fill well in the cell (but this depends on battery size and design; check with your manufacturer). Do not overfill and never add acid to a battery except to replace spilled liquid. Only use distilled or de-ionized water to top off non-sealed batteries.

Electrical Connections—Make sure all battery terminals and cable connections are clean and tight. Dirt on a battery's surface can creep and lead to discharge and corrosion. The application of a thin layer of petroleum jelly or similar anti-corrosion product to the terminals can help reduce corrosion.

Cleanliness—Batteries behave best when they're kept clean and dry. Avoid spilling oil or grease onto the top of the battery. To remove dirt or moisture, wash with a solution of bicarbonate of soda and water but do not allow any of the solution to get inside the battery. Rinse afterward with clear water and dry with a clean cloth. Ensure vent plugs are in place at all times. A clean battery is a happy battery.

Temperature—Batteries need to be kept at reasonable temperatures to perform well and live long lives. A good rule of thumb is to try to keep batteries at temperatures you would find comfortable. In other words, not too hot and not too cold. If you live in a cold climate, a battery heater system might be in order. Lead acid batteries can lose up to half of their capacity at 32 degrees Fahrenheit. Usually battery heaters take the form of mat-style heaters that are placed under the batteries. Insulated battery boxes are also a good idea in cold weather.

Note: Never attempt to charge a frozen battery. Warm temperatures can also affect performance. Usually, the rated capacity is not changed, but the battery life itself can be reduced. If you operate your EV in a hot climate, make sure the batteries are well ventilated. You should also avoid charging batteries if the temperature is above 120 degrees Fahrenheit.

Charging—Make sure your charger is sized correctly for the battery bank voltage. If the charger is programmable, make sure that it's set up correctly for the type of battery you're connecting to; the charge requirements for flooded batteries is different than that for gel and AGM-types, for instance. When you buy a new set of batteries and charge them for the first time, make sure you fully charge them up. After charging, let the batteries sit for approximately 24 hours and/or apply a small load to the batteries to remove any surface charge. Afterward, take hydrometer readings after charging and record the data for future comparisons. Note that most batteries will not reach full capacity until cycled 10–50 times, or more. A brand-new battery can have a capacity of about 5–10% less than the rated capacity.

Discharging—Batteries need to be cycled; exercise is good for them, but not too much or too little. Lead-acids provide the longest life when they are worked to about 50% of their capacity at moderate loads. Both under- and over-discharging can cause problems and shorten the useful life of the battery. Unlike some other types of batteries, lead-acids do not have a memory. The old wives' tale that they should be fully discharged to avoid this memory effect is false and will just lead to early battery failure.

Recharging—Batteries need to be recharged as soon as possible after use. Batteries left sitting in an undercharged state can develop something called sulfation. On the other hand, don't go overboard and overcharge a set of batteries, either, as this can also shorten the lifespan of a battery. Also, keep vent caps on during charging to minimize fluid loss. Before charging, ensure that the internal electrolyte fluid levels are correct.

Equalizing—The life of flooded-type lead-acid battery can be extended if they are subjected

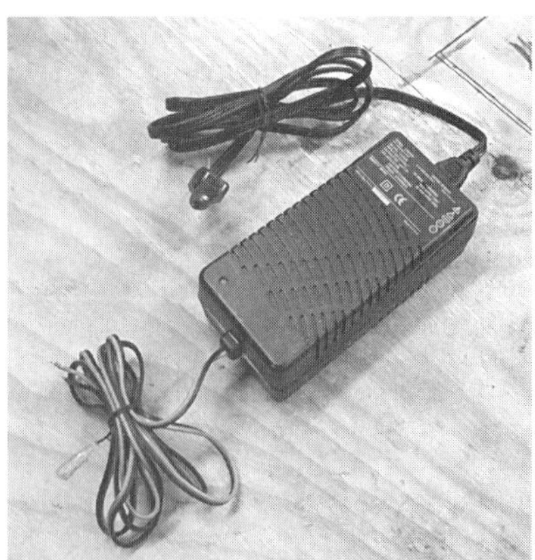

A Soneil 72-volt battery charger. Courtesy Tony Helmholdt.

The same Soneil 72-volt battery charger mounted on an electric motorcycle allows for opportunity charging wherever the bike is parked. Note the removable power cord. Courtesy Tony Helmholdt.

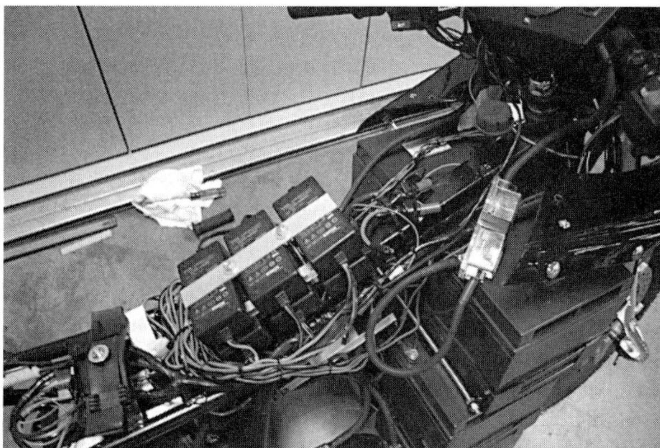

Another motorcycle charger application. This owner elected to install six separate chargers, one for each of the six onboard batteries. This allows precise control over the charging profiles for each unit. Courtesy Lennon Rodgers.

This clever recharge connector is designed to be placed in the original gasoline filler hole on the side of a car or truck. Courtesy Electric Car Company of Utah.

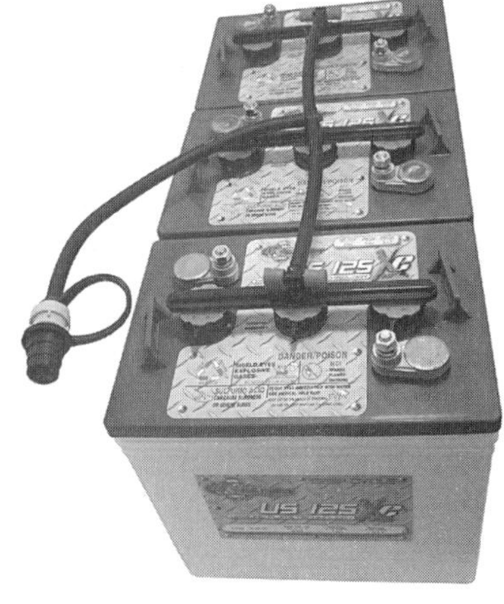

If you need to add water to all the cells of a battery bank, you can use a system like this one. Be careful however, as individual cells may require more or less water than their neighboring cells. Courtesy U.S. Battery.

periodically to an equalizing overcharge charge. Typically, this type of charge is roughly 10% higher than the normal full voltage charge and is applied for 2–10 hours or more. This step ensures that all the cells are equally charged and that gas bubbles are able to fully mix the electrolyte. Equalizing is recommended only when batteries have either a low specific gravity or a wide range of specific gravities between cells after fully charging the battery. Note that you should never equalize-charge gel or AGM-type batteries.

Age—As batteries age, their maintenance requirements change, including longer charging times. Usually older batteries also need to be watered more often. Capacity can begin to fall off with advanced age, too.

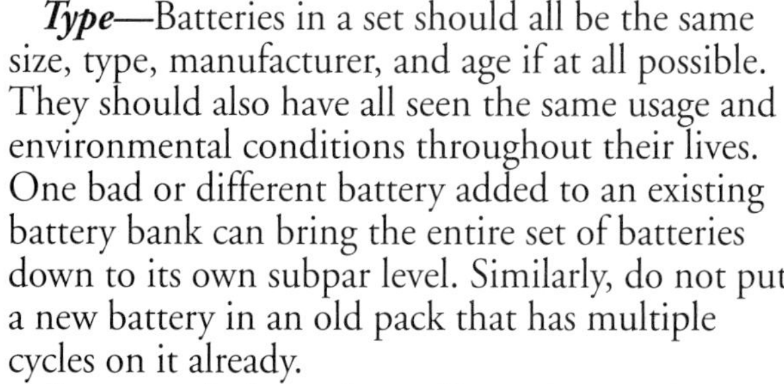

Type—Batteries in a set should all be the same size, type, manufacturer, and age if at all possible. They should also have all seen the same usage and environmental conditions throughout their lives. One bad or different battery added to an existing battery bank can bring the entire set of batteries down to its own subpar level. Similarly, do not put a new battery in an old pack that has multiple cycles on it already.

Storage—A charged set of batteries can sit unused for short periods without needing to be attached to a continuous trickle charger. Just hook them up occasionally so they stay near full charge. Always charge a battery before putting it into storage, which should take place only in cool, dry areas protected from the elements. Avoid direct exposure to heat sources. Batteries gradually self-discharge during long periods of storage. Every 4–6 weeks, you should monitor the specific gravity and/or voltage levels. If the state of charge falls below 70%, the batteries should be recharged. If the ambient temperature is high (e.g., greater than 90F) monitor the specific gravity of the batteries more frequently. When the battery is removed from storage, you should recharge it prior to use.

Wiring up all the parts and components of an EV together requires good planning and patience. It also requires you to purchase the correct wire type and gauge and know how to make safe and reliable connections. Do it right and you will have thousands of trouble-free miles of EV ownership. Do it wrong, however, and you can count on frustration and poor, unreliable performance. Worse, you can create a vehicle that is unsafe and prone to electrical shorts, shocks, and fires. Courtesy Fred Weber.

Chapter 8

Wires, Switches, and Tools

"When you look at the inner workings of electrical things, you see wires. Until current passes through them, there will be no light."
—Mother Teresa

If the motor, controller, and battery of an EV are akin to a human body's heart, brain, and lungs, then the wires, fuses, and switches that connect these vital components would be equivalent to the body's veins and arteries. Without all the ubiquitous under hood connectors and wires, nothing would work as intended. More important, if the wire connections are unreliable or intermittent, the result would be, at best, a vehicle that is unreliable to drive. At worst, the EV can be unsafe to operate.

In this chapter, we'll go through some of the key components in your car's wiring, starting with the wire itself. Then we'll discuss switches, fuses, and connectors, and then see how it all fits together. We will also look at some of the more common and useful tools you will use when wiring up an EV project vehicle.

Wire

Insulated copper wire has to perform one simple task—provide a conductive path for electricity to flow from point A to point B—so it would seem that not much can be said about its design and implementation. What could be simpler than a piece of copper wire, right? Wrong. Electrical wire is available in a myriad of types, sizes, and materials. Choosing the correct wire size and type to connect all the point As to point Bs is critical to the success of an EV project. The chief concern of selecting the wrong wire is overheating (and fire).

Wire overheating occurs when too much current passes through a wire. Said another way, if the wrong wire type or size is selected for the amount of current the circuit requires, the copper will heat up (due to its own internal resistance) and melt its insulation. This can lead to shorting the circuit (which can lead to even higher current flows), or it can cause a fire by heating up flammable components near the path of the wire. In either case the results can be catastrophic.

The amount of current a wire can safely handle depends on a number of factors. The most important of these are its gauge, length, composition, and how it is packaged, or grouped together throughout the vehicle.

Gauge—The primary means by which wire is classified is its gauge, which is a measurement of how large the wire is, either in diameter or cross-sectional area. There are a number of wire gauge classifications used in different industries, but the most common is the American Wire Gauge, or AWG specification.

The AWG gauge system originated with the number of drawing operations used to produce a given gauge of wire. Very fine wire (e.g., 24-gauge) required more passes through a set of drawing dies than did an 8-gauge wire. Consequently, the larger the gauge number, the smaller the wire diameter; i.e., a 24-gauge wire is smaller in diameter than a 20-gauge wire. The smallest AWG wire is 40-gauge, which has a nominal diameter of just 0.003". Wires get progressively

Electrical wire comes in a wide variety of sizes, shapes, types, and gauges, from large "welding"-type power cables to small diameter stranded wire.

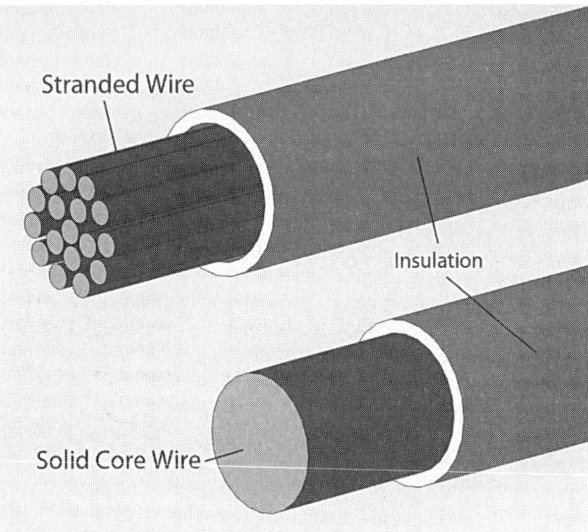

Electrical wire comes in two major types: solid core and stranded. For most automotive applications, stranded wire is preferred. This is because it is more flexible and forgiving in the high vibration environment of a car driving on rough roads.

larger in diameter as the gauge number drops, with gauges going all the way down to zero, which is often written 1/0, and pronounced "one aught." Even larger than this is 00 (often written as 2/0, and pronounced "two aught"), and so on. The diameter of a 4/0 wire is almost half an inch.

The lower the gauge number, the lower the resistance of the wire. Said another way, the lower the gauge number, the higher the current carrying capacity of the wire is.

Length—Any given length of wire has a certain amount of resistance; i.e., each type of wire has a certain amount of resistance per linear foot. The longer the wire, the larger the overall resistance it has. A short piece of wire may safely handle a 10 amp electric current, but if its length is increased significantly, the wire may overheat when subjected to the same 10 amps. When selecting a wire for a long run from the front a car to the rear, it's often better to err on the side of selecting a wire gauge that is one size larger than nominally required.

AWG Gauge	Conductor Diameter, Inches	Resistance, Ohms/1000ft
OOOO	0.460	0.05
OOO	0.410	0.06
OO	0.365	0.08
O	0.325	0.10
1	0.289	0.12
2	0.258	0.16
3	0.229	0.20
4	0.204	0.25
5	0.182	0.31
6	0.162	0.40
7	0.144	0.50
8	0.129	0.63
9	0.114	0.79
10	0.102	1.00
11	0.091	1.26
12	0.081	1.59
13	0.072	2.00
14	0.064	2.53
15	0.057	3.18
16	0.051	4.02
17	0.045	5.06
18	0.040	6.39
19	0.036	8.05
20	0.032	10.15
21	0.029	12.80
22	0.025	16.14
23	0.023	20.36
24	0.020	25.67
25	0.018	32.37
26	0.016	40.81
27	0.014	51.47
28	0.013	64.90
29	0.011	81.83
30	0.010	103.20
31	0.009	130.10
32	0.008	164.10
33	0.007	206.90
34	0.006	260.90
35	0.006	329.00
36	0.005	414.80
37	0.005	523.10
38	0.004	659.60
39	0.004	831.80
40	0.003	1049.00

The size of electrical wire is specified by its gauge. The larger the AWG gauge number, the smaller the wire diameter and, hence, the lower the wire's capacity for carrying electrical current.

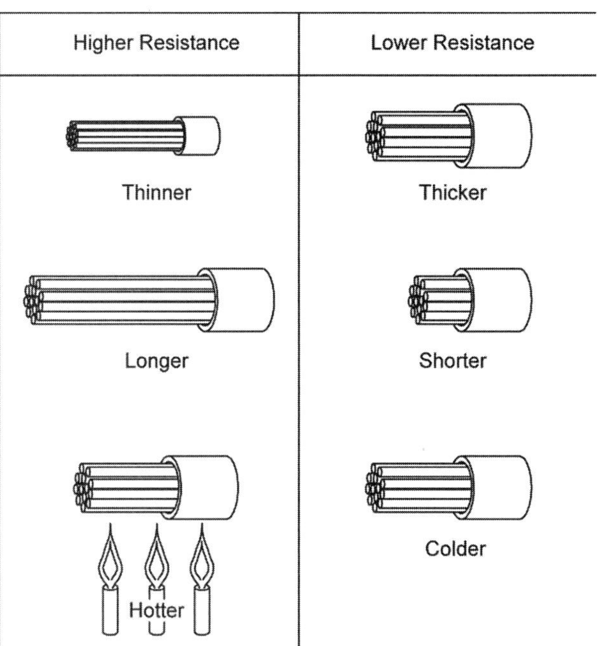

The resistance of an electrical wire depends on a number of things. Generally speaking, the longer the wire and/or the smaller its diameter, the higher the overall resistance. Temperature also affects electrical resistivity, as does the wire material itself, as well as the physical condition of the wire.

Do not be tempted to use inexpensive, low quality lugs and connectors. A little bit of extra money will go a long way toward ensuring a reliable operating vehicle. These high quality tinned copper closed-end lugs provide excellent corrosion resistance and high electrical conductivity. Courtesy EVSource.com.

A high quality terminal lug connector for use with high amperage 4/0 power cable. Courtesy EVSource.com.

Some EV builders recommend the use of anti-corrosion paste inside crimped connector joints and between battery terminals and connector lugs.

Copper or steel "buss bars" are sometimes used in place of electrical wire or cable in high amperage circuits to connect components that are physically close to one another. Note the use of insulating electrical tape wrapped around the unit, which leaves just the ends exposed. Courtesy Tony Helmholdt.

Composition—Copper wire is available in two primary types: solid and stranded. Solid wire, which is also sometimes called solid core or single-strand wire, consists of a single piece of metallic wire, usually copper. In contrast, stranded wire is comprised of a number of smaller gauge wires bundled together to make a larger conductor.

Generally speaking, for EV and other automotive applications, stranded wire is preferred. It is more expensive than solid wire but is also more flexible, kink-resistance, break-resistance, and better suited to the harsh, vibration-prone environment of a vehicle driven on the road. The finer the strands (i.e., the larger the number of strand in a wire), the lower the resistance and the more current that can be carried.

Grouping—The way a wire is grouped with others can affect how well it dissipates heat, which is primarily accomplished by way of convective heat transfer to the surrounding air. If a wire is in the center of a group of twenty other high-current wires, it won't be able to dissipate heat and, consequently, will not be able to carry as much current as if it were placed alone in free air.

Switches, Relays, and Contactors

To control the flow of electricity in your EV circuits, a number of switches are required. A switch is simply an electrical device that can break a circuit, thereby disrupting the flow of electrical current. For example, a standard room light switch simply "closes" (i.e., makes, or completes) or "opens" (i.e., breaks, or disrupts) the flow of electricity to the room lamp.

All switches have a set of contacts, which can be either switched open or closed. A switch is said to be closed when electrical current can flow, and, conversely is said to be open when the circuit is interrupted. The internal arrangement of the contacts can come in many different types and configurations. The terms *pole* and *throw* are often used to describe these contact variations. A pole is a set of contacts and terminals that, when connected, form a single circuit path. A throw is one of two

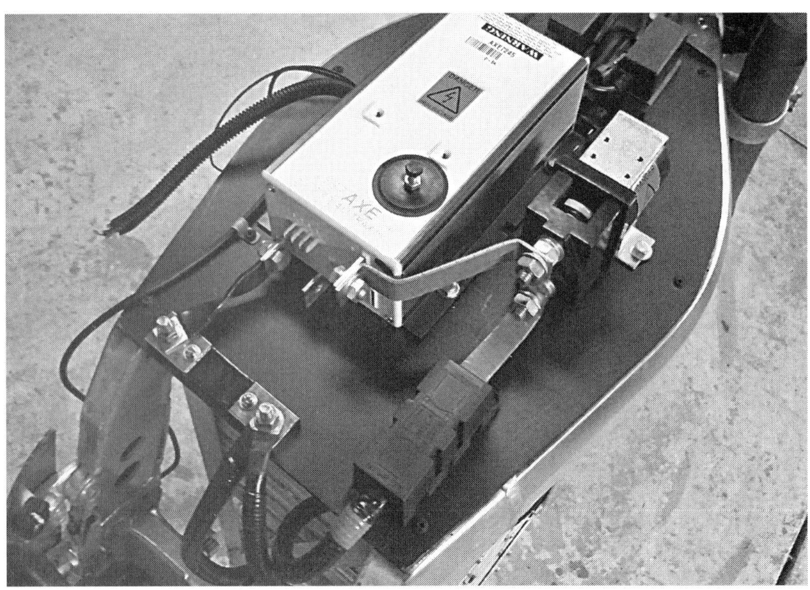

The owner of this electric motorcycle has used buss bars to connect the controller, main fuse, main contactor, and shunt resistor together. Note the short, direct connections between the components, which helps minimize resistive power losses. Careful and thoughtful planning goes into the creation of a clean installation like this. Courtesy Tony Helmholdt.

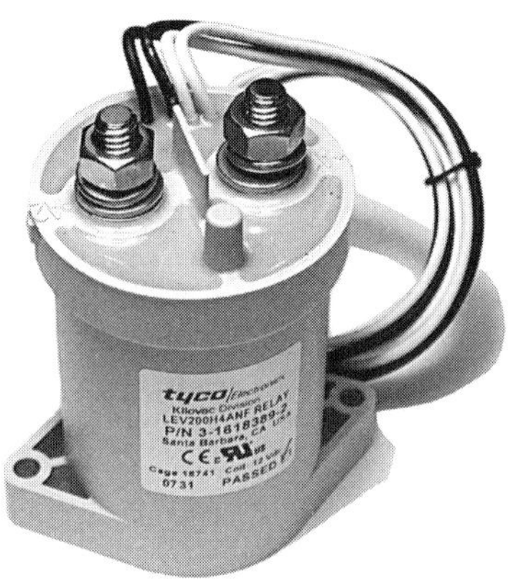

A hermetically sealed power contactor, like this Tyco Kilovac 500 amp 320 volt LEV200 unit can be used in harsh, explosive-prone spaces such as battery boxes. Typical usage at high currents requires the contactor to be heat-sinked. Courtesy EVSource.com.

To prevent accidental shocks, remotely actuated high power contactors are used to switch on and off power in electric vehicles. This particular contactor is rated at 250 continuous amps and 96 volts. Activation is via low power 12-volt input. Courtesy EVSource.com.

When using power contactors, always ensure that the unit you purchase will be operating with its power ratings. Courtesy Electric Car Company of Utah.

different positions that the switch can be placed in. A single-throw switch has one position that closes the contacts, while a double-throw switch has two positions. These terms are used in abbreviations for the type of switch configurations. The simple on-off room light switch mentioned above would be an example of a single pole, single throw (SPST) switch.

Switches can be either manually or electronically operated. Manual switches are available in a variety of styles and types, including toggle, momentary, push button, rotary, rocker, and so on. There are also removable key-switches that can be used to fully disable an electrical circuit from use.

Electrically operated switches are used when the user needs to remotely operate the switch, either for convenience or safety reasons. For example, an electrically operated remote switch is typically used to switch on and off the main electrical power circuit coming from the battery to the controller and/or to the motor itself. Keeping these high-power, potentially dangerous circuits outside of the cabin and away from the driver and passengers is very important for the design of a safe EV.

Electrically operated switches fall into two major categories: relays and contactors. Strictly speaking, relays and contactors are both very similar devices, with the terms often interchanged. The distinction usually is based on the amount of electricity the switch is designed to handle; relays are normally used in low-power applications and contactors are used in high-power circuits.

Fuses, Fusible Links, and Circuit Breakers

Every circuit in your EV, from the major power, battery, and motor circuits, down to the small accessory and component lines, should be fitted with some type of over-current protection device, such as a fuse, fusible link, or circuit breaker. Without such a safety device inline, a power surge or a short circuit could result in equipment burning out and damage to circuit boards, components, and other electronic devices. More important, a power surge or short circuit could cause a fire or further damage to the other electrical systems within the vehicle.

A short circuit is an abnormality in an electrical circuit that allows current to travel along a different path from the one originally intended. For instance,

If your controller doesn't have reversing capabilities built in, a reversing contactor set, like this SW202A unit, can be used. Courtesy EVSource.com.

Do not skimp when it comes to over-current protection. Shown here are quality Ferraz-Shawmut fuses. The upper fuse is a fast-acting high current fuse for applications such as the main battery feed to the controller. The lower fuse is a low-current device suitable for lower power circuits. Courtesy EVSource.com.

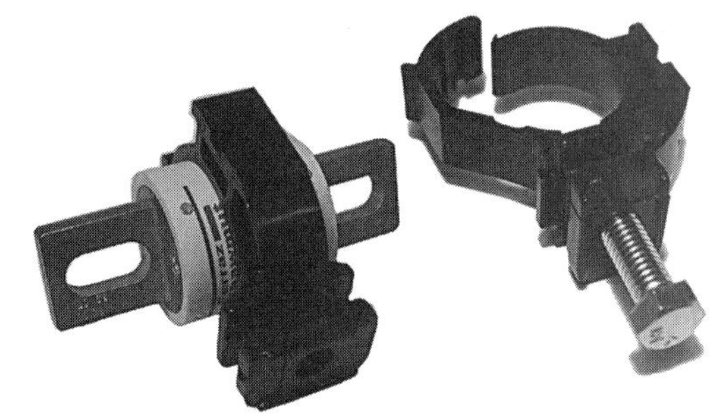

Fuse holders are available in a variety of types and sizes. Shown here is a Ferraz-Shawmut fuse clamp for securing a high-current fuse in place. Courtesy EVSource.com.

if the insulation around a wire is damaged, and the inner conductor is allowed to contact a metallic portion of the vehicle's chassis, the electrical current may "short out" and take the path of least resistance into the chassis and back to the battery bank. This can result in extremely high current flows from and to your battery bank, through components, the chassis, and even you. This runaway current can cause a fire or explosion if it's not stopped. For this reason, multiple series-wired over-current, or emergency disconnect devices are recommended for the main power systems. Ideally, you should have at least two, and these should be different types. Your choices include fuses, fusible links, and/or circuit breakers. The reason you want multiple, different types is in the event that one of them fails, the other(s) can act as backups and interrupt the circuit. Smaller, less critical circuits can be served with a single, high-quality over-current protection device.

Fuses

A safety fuse is an over-current protection device that consists of a metallic element (i.e., a metal strip or wire made from aluminum, copper, silver, zinc or other alloys). This element has a relatively small cross-sectional area when compared to the circuit wiring it is attached to. The element is mounted between a pair of electrical terminals and is usually enclosed in a housing that is both non-conductive and fireproof.

In electrical circuits, fuses are usually wired in series (i.e., in line) so that all the electrical current in the circuit has to flow through the fuse. When the current flows through the fuse, the resistance of its small cross-sectional area creates heat. If the current is high enough, the heat raises the temperature to a point where the metallic element melts, or "blows." This causes a break, or open, in the circuit, which prevents any current from flowing through the circuit. This in turn prevents any further damage to equipment downstream of the fuse. After the fault that caused the over-current to occur is repaired, the fuse itself needs to be discarded and replaced with a new fuse. Unlike circuit breakers, fuses are non-reusable.

Fuses need to be sized and rated for the current, voltage, and time characteristics appropriate to the application. Fuses are available in "fast-blow," "slow-blow," and "time-delay" variants. When selecting a fuse, you must ensure that small, normal spikes or surges in current expected in your motor circuit do not inadvertently blow the fuse. That said, you don't want your fuses to take too long to disconnect the circuit, either. It's best to discuss your application and component selections with the manufacturers or EV suppliers to ensure you use the appropriate fuse.

Fusible Links

A fusible link is a short piece of special wire designed to melt if too much current travels through it. The link itself is often encased in a special high-temperature insulated sleeve that is fireproof.

Like fuses, fusible links are sacrificial devices. After the fault that caused the over-current event is repaired, the fusible link needs to be discarded and replaced with a new link. This typically means cutting out the old link and soldering in a new one.

Fusible links are often used in locations where

Byrne's Law

"In any electrical circuit, appliances and wiring will burn out to protect the fuses." —Robert Byrne

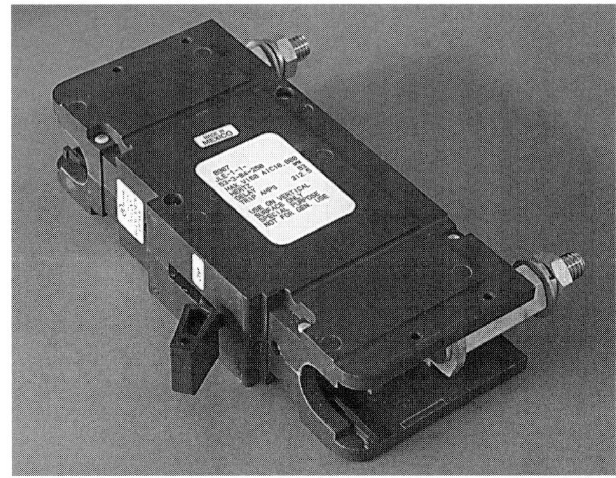

An alternative to fuses is a circuit breaker, like the Airpax unit shown here. The chief advantage of a circuit breaker is that it's reusable after it's been tripped. The downside is that it is more costly than an equivalent fuse. Some EV experts suggest that both fuses and circuit breaker be employed in particularly critical circuits; if one unit fails (e.g., the circuit breaker), the other (e.g., a fuse with a slightly higher amperage rating) can provide backup over-current protection. Courtesy KTA Services.

there isn't room for a fuse block or in applications where high current spikes are frequent (and normal) and would blow a traditional fuse. Fusible links are generally more forgiving than a fuse, with the ability to handle temporary surges without blowing. For this reason, they are often suitable in high-current power lines. This is also the reason that substituting a fusible link in a circuit in place of a fuse may not be recommended; depending on the circuit and type of protection required, the fast acting response of a traditional fuse may be desired—not the slower reaction of a fusible link.

Unlike fuses, fusible links are not normally rated in units of amps. Instead, fusible links are often sold by their overall size, or wire gauge. The typical rule of thumb is to select a fusible link four American wire gauges smaller than the nominal wire size in the circuit that is being protected. For example, an electrical circuit that uses 14-gauge wiring may employ an 18-gauge fusible link. Unfortunately, this rule of thumb should be considered just a guideline. For selection of a fusible link, it's recommended that a qualified electrical engineer, who has full knowledge of the circuit protection requirements, installation and operating conditions, and safety needs, help select fusible links for a given application.

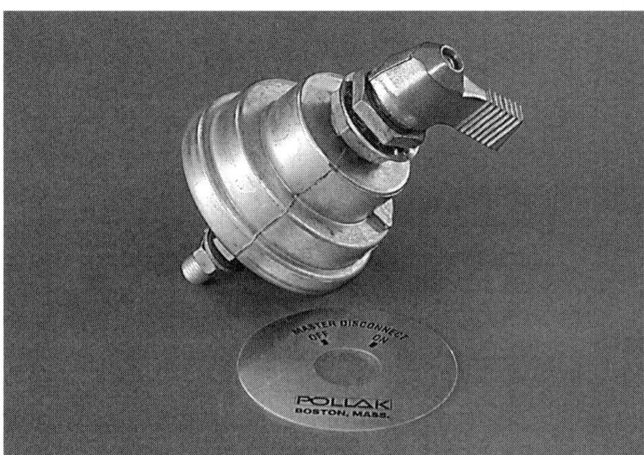

For complete protection during maintenance operations, consider installing a master power disconnect switch in your EV, such as this Pollak unit. Courtesy KTA Services.

Circuit Breakers

A circuit breaker is essentially an automatically operated electrical switch that opens whenever an overload or short circuit is detected.

Unlike fuses and fusible links, which operate once and then need to be replaced, circuit breakers can be reset to resume normal operation. Resetting can be either via manual operation or can be automatic, depending on the type of circuit breaker.

Some circuit breakers operate on an electro-magnet principle. The internal electrical contact switch is connected to an electromagnet solenoid. As current flows through the device, an electromagnet is energized; the more current, the higher the electromagnetic force generated. Above a certain level of current, the solenoid is strong enough to open the switch, thereby stopping the flow of electricity in the circuit.

Other circuit breakers utilize bimetallic strips of metal attached to a release mechanism on the internal electrical switch. As the electrical current rises, the resistance of the metals causes a rise in temperature. Because the two metals used in the bimetallic strip have different coefficients of thermal expansion (i.e., they expand different amounts for a given temperature rise), the strip bends. This bending causes the release mechanism to trip, thereby opening the switch and cutting power.

Some circuit breakers employ both an electro-magnetic solenoid and bimetallic strips. The electromagnet is used to trip the circuit under high surges of current, such as those seen during a short circuit event. The bimetallic strip is used for over-temperature situations and slower-acting rises in current. Still other types of circuit breakers employ sophisticated electronics to constantly monitor current levels, instantaneously tripping the internal switch if the amperage is too large.

When shopping for a circuit breaker, make sure you use units rated for use in automobiles, where they will be subjected to relatively large ambient temperature swings, as well as to significant vibrations when the vehicle is underway on the road.

Basic Tools: Crimpers, Cutters, Strippers, Heat Gun

Now it's time we look at some of the more common tools you will probably use during the conversion of an ICE-powered vehicle to electric power. This is not meant to be an exhaustive list, nor is it a compilation of recommendations for which tool is better or more useful than another. Instead, we're simply going to talk a little about some of the more useful tools that will make the construction of an EV easier and safer. The old adage "use the right tool for a job" is entirely applicable to EV construction. In other words, investing in a set of proper tools will go a long way toward ensuring you are able to safely build a reliable electric vehicle—and enjoy the conversion process along the way. Trying to make do with substandard tools is a certain path to taking shortcuts and/or just making the project a miserable experience.

Wire Strippers—One of the first tools you will need to learn to be proficient with is a wire stripper. The average EV project will have hundreds of feet of wire, along with dozens, if not hundreds of connections. Accurately and safely stripping the ends of wiring for making all these connections can be performed with a razor and a steady hand, but it's not recommended. Instead, it is suggested you invest in a quality wire stripper.

Wire strippers come in two basic types: manual and automatic. A manual stripper is simply a device that looks similar to a pair of pliers, but instead of gripping surfaces, the stripper utilizes a pair of opposing blades. A center notch in each blade holds the wire and creates a pivot point about which the stripper is rotated. The insulation is literally cut around the copper wire core while pressure is applied. Once the insulation is fully cut, it can be simply pulled off the end of the wire. Some manual strippers have a series of different size notches that allow easier cutting of different size wire insulations.

Automatic strippers are more expensive than manual, but quality units make insulation removal extremely easy. An automatic stripper simultaneously grips the wire from one side, while cutting and removing the insulation from the other side. The chief advantage of an automatic stripper is its ease of use; even novices can quickly and accurately remove the insulation from the ends of a variety of different wire sizes. The main disadvantage, however, is that the wires being stripped have to fall into a range of sizes. If a wire is too small in diameter for the stripper, the copper conductor itself can be damaged. Conversely, wires that are too large in diameter simply won't fit into the jaws of the stripper.

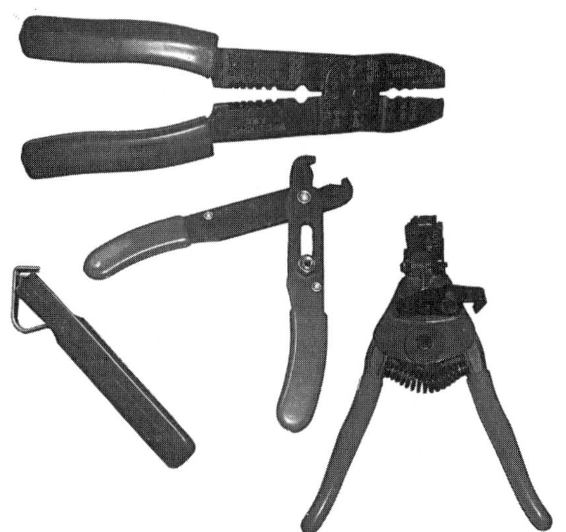

A variety of different cable strippers are available for purchase. It doesn't really matter which type you buy, as long as you spend ample time practicing prior to stripping actual wires in your EV.

Soldering Irons—One of the two major ways wires can be joined together is via soldering, and the tool used to perform this task is known as a soldering iron. (The other major means of joining wires is by way of crimping, which we will discuss a little later on.)

A soldering iron is a device comprised of a small metal rod or bar (called the *iron*) that has an internal heating element. The end of the bar is typically formed into a pointed tip, and the other end has an insulated handle. The internal heating is usually provided by way of an internal resistive element that warms up when electrical current is passed through it. Other types of soldering irons use combustion of a gas (such as butane or propane) to heat the iron.

When in use, a soldering iron is switched on to heat up and is then left on in the heated state on a stand or wire support frame. Some soldering irons can heat up (and also cool down) in just a few seconds, while others require several minutes to come up to operating temperature. The support stand also usually comes equipped with a sponge and flux pot that is used to periodically clean the tip of the iron. Higher end soldering irons come with soldering station stands that have an adjustment dial that can be used to set the exact

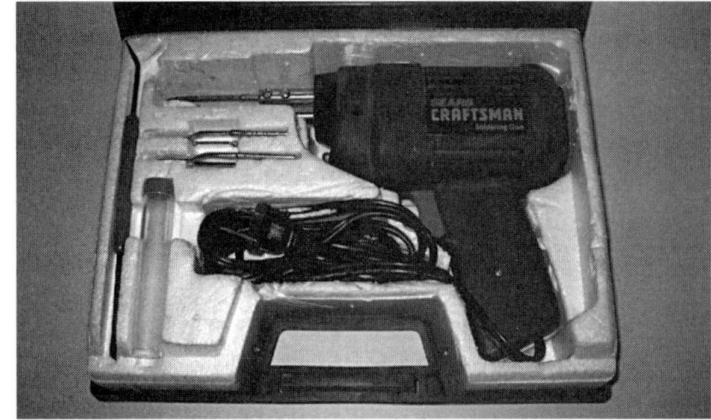

When building an EV, you need to solder wires to connectors and to other wires. Invest in a quality soldering iron and your life will be made easier. Skimp on the iron, and the job of wiring your EV will not be fun.

Cutting large diameter welding and power cables can be difficult without the right tool. This cable cutter is capable of severing up to 6/0 power cable. Courtesy EVSource.com.

A hammer-type cable connection crimper is used to make crimps on wire sizes from 8 gauge up to 4/0 cable. The cable and connector are inserted, and then the crimp is made by either striking with a hammer, or inserting the unit into a vise or press. Always wear safety glasses when using any type of striking tool like this. Courtesy EVSource.com.

temperature of the iron tip at a proscribed level.

Another feature to consider when shopping for a soldering iron is the tip type. Older and less expensive irons are often fitted with bare copper tips, which must be shaped with a file or emery paper and is subject to pitting and erosion of the working surface. Higher end soldering irons use plated copper tips, which is more erosion resistant and generally have longer operational life spans.

Crimping Pliers—Besides soldering, the second major method of joining two wires together (as well as attaching a wire to a connector or terminal) is crimping. There are a large number of different crimp terminals and connectors available on the market, but they all basically rely on the same process: namely, the mechanical deformation of a device onto the wire.

Crimp connectors are not normally used on solid core wire; that is, crimping is almost exclusively used on stranded-type wire. To crimp a connector onto a wire, the end of the wire is usually (but not always) stripped to reveal the copper strands. The wire is then inserted into the connector and a tool, called crimping pliers, is used to compress and deform the connector around and into the wire strands. A properly crimped connection is actually gas-tight and will provide a very low resistance path for electricity to flow from wire to connector. Some crimp-type connectors include a secondary crimp surface used to grip the insulated portion of the wire (i.e., in addition to the primary crimp that takes place on the stripped copper wire).

The chief advantages of crimping over soldering are that the process is very quick and doesn't put any heat into the wire, which, if not controlled carefully can melt the insulation. If done properly, the connection can be every bit as good as a soldered joint, plus can have slightly better mechanical strain relief characteristics. Note the use of the phrase "if done properly." If a crimped connector isn't deformed correctly, the connection may not perform properly, with either high resistance or intermittent opening of the electrical circuit.

The key to getting a properly crimped connection is to use a high-quality set of crimping pliers designed for the specific type of crimp connectors used on the wire. Some amateur builders forgo the use of crimping pliers and instead simply use side cutters or other pinch-type pliers to deform the metal. While this can be made to work, the uncertainty of getting a perfect connection is high, especially when you have to make 50-100 or more such connections on an EV circuit. For this reason alone it is highly recommended that you purchase a high-quality set of crimping pliers and learn how to use them. Purchase a few extra feet of wire and a dozen or so extra connectors to practice with prior to using them on the final wiring under the hood.

Heat Guns—Many of the soldered and crimped connections you're going to make when constructing an EV will require heat shrink tubing covers. You may also need to de-solder wire and circuit connections. The tool with which you can perform these tasks is called a heat gun.

A heat gun looks (at least superficially) like a typical handheld hair dryer you might have in bathroom. The difference, however, is that heat

Another useful tool is a proper heat gun, which is used for applying heat shrink tubing. Courtesy EVSource.com.

guns can operate at much higher output temperatures. Normal output temperature of the airstream from a quality heat gun can be 300 to 1000-degrees Fahrenheit.

When shopping for a heat gun, look for a brand name unit that has a decent warranty, as the cheaper, off-brand models are known for burning out after just a few uses. Also look for a unit that has both temperature and airflow control adjustments.

Multimeters—While not strictly necessary for the construction of an EV, the addition of a quality multimeter (also known as a multi-tester, volt-ohm meter, or just VOM) is recommended. Being able to measure resistance, voltage, and current in a circuit can greatly speed up troubleshooting tasks.

Multimeters are available in a very wide range of makes, models, and prices. Low-level units can be purchased used online for as little as $25 or less. Higher end multimeters designed for professional use can cost thousands. For the average EV constructor, however, a basic digital unit that can be switched between volts, ohms, and amps is probably sufficient. The ability to automatically display the levels being tested is a nice feature. Also, unless the price is extremely attractive, it's recommended that you forgo purchasing an older analog type tester and instead opt for a modern digital readout unit.

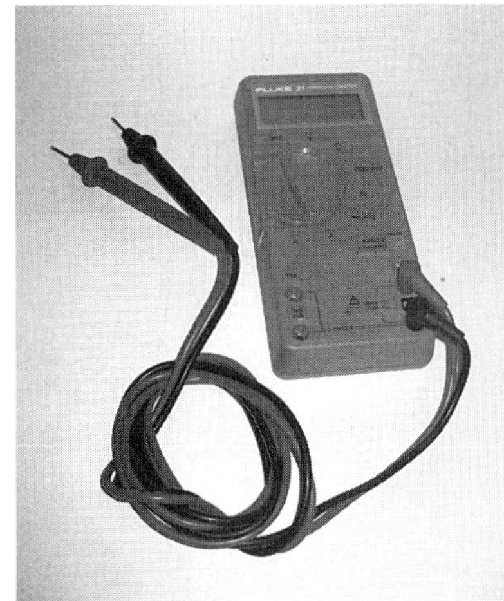

Testing and troubleshooting electrical circuits is made relatively easy with a low-cost digital multimeter, like the Fluke unit shown here. Resistance, current, and voltage are all easily measured with the probe wires.

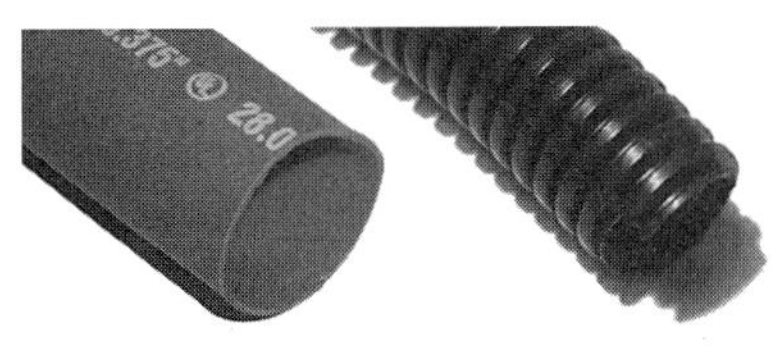

Once you've laid out your wiring and connections, it is recommended that you cover the wires with a loom and heat-shrink all connections where possible. Besides protecting the wirings from chafing and accidental shorts, this type of treatment creates a professional-appearing, tidy engine bay. Courtesy EVSource.com.

An electric motorcycle halfway through its wiring layout. Courtesy Tony Helmholdt.

Companies like the Electric Car Company of Utah sell professionally wired boxes that come with main power contactors, controllers, cooling fans, and other equipment pre-installed. To install, this entire safety-rated box is bolted into a waiting engine bay. Courtesy Electric Car Company of Utah.

Use only high-quality parts and equipment when building an EV. This philosophy should extend to even simple, commonplace items like pressure-sensitive electrical tape. Thick, quality tape will not "unstick" with time and will provide years of reliable insulating capability.

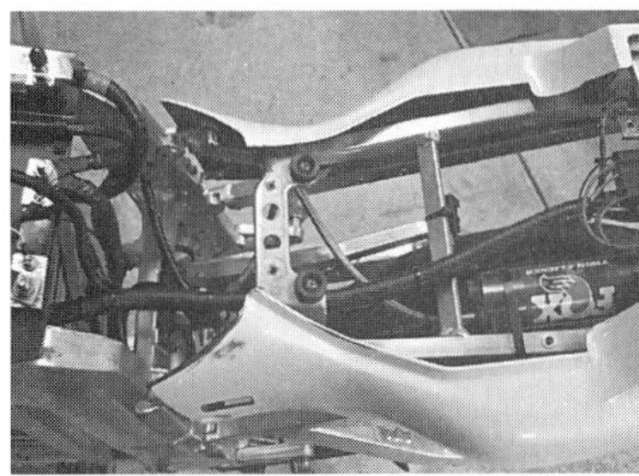

The same electric motorcycle after heat shrink and split loom is applied. The difference is subtle but important. Note the use of zip ties and wire-loom hold straps to ensure the wires do not chafe and short out. You must be careful to leave enough slack in wires so they are never subjected to tension or continuous cyclic bending. Courtesy Tony Helmholdt.

Chapter 9

Accessories

"If you go to an all-electric system.... You only use power when you need it. You only use air-conditioning when you run the compressor. You only use steering power when you turn the wheel." —Robert Schumacher

A DC-to-DC converter is used to step down the high battery bank voltage of an EV to 12 volts for use in all of the secondary and auxiliary electrical circuits, such as the lights, horn, radio and wipe motors. Shown here is an IOTA power converter that can reduce 108–190 volts to 12 volts. Courtesy EVSource.com.

In this chapter we will look at some of the accessories and auxiliary components you might consider including in your EV project vehicle. None of these items are strictly necessary when planning a conversion project, but they do offer certain modern conveniences and amenities that make owning and operating an EV much more enjoyable.

DC-to-DC Converters

Even the most basic electric vehicles are comprised of more than just a battery bank, electric motor, and controller. Lights, turn signals, horn, windshield wipers—all of these things are required on a car or truck conversion that will be driven on the road. More advanced vehicles also have modern conveniences, such as power windows, door locks, fans, and stereos. All of these items require electric power to operate, and that power is usually supplied in the form of a 12-volt auxiliary power circuit.

There are several ways to supply the 12 volts required to operate these accessories and auxiliary equipment. A few backyard-engineered EVs have been built that utilize an alternator that is belt-driven off the electric motor in much the same way as an ICE-powered vehicle operates. This type of system can be made to work, but it's not recommended. Inefficiencies in the belt, bearings, and alternator itself make this Rube Goldberg approach a less than ideal solution. Also, the system only works when the motor is turning, and the motors used in EVs come to complete halts whenever stopped at a light or stop sign.

Other low-budget EVs get away with simply installing a large, deep-cycle 12-volt battery to run the accessories directly. The battery is recharged (via a separate charger) whenever the main battery bank is being recharged. This is another simple, but non-ideal solution, as it often results in weak lighting and poor accessory performance.

You might also be tempted just to tap into one of the 12-volt batteries in your main battery bank (or tap into two of the six volt batteries if you're running a system based on 6V units). The main problem with this approach is that it can unbalance the battery bank, causing a few of the batteries to drain more quickly than the rest, thereby dragging down the entire bank. This also can cause safety issues if the accessories use the OEM method of the chassis as the ground return path; i.e., the main battery bank is now connected to the chassis, which can lead to dangerous shorts and the possibility of electrocution.

So, if none of the aforementioned methods are recommended, what's the preferred solution for providing a reliable source of 12-volt power to lights, wipers, and radio? The answer is something called a DC-to-DC converter.

A DC-to-DC converter is an electronic power supply that

is connected to the main battery bank, which takes the high voltage and steps it down to 12 volts so that it can be used to power the accessories. Because the 12 volts are essentially isolated from the main battery bank, the accessories can use the chassis as the ground return path. Because the unit uses modern electronics (e.g., pulse width modulation), the output voltage is held steady at 12–14 volts, regardless of the state of charge of the main battery bank. This ensures that lights and accessories function properly at all times. Also, whenever the main battery bank is charged, this auxiliary circuit is effectively recharged as well.

Some DC-to-DC converters can be wired directly into the accessory circuits, while others use a small 12-volt buffer battery between the DC-to-DC converter and the accessory circuits. There are minor pros and cons of each approach, but neither system is significantly better than the other. For instance, using a buffer battery means extra weight and complexity, plus the battery needs maintenance. On the other hand, if the DC-to-DC converter ever fails, the buffer battery can provide emergency lights for driving home for repairs.

Some enterprising DIYers have successfully built their own DC-to-DC converters from power supplies available at electronic and computer stores. Unless you're an expert in electronics, however, this is not recommended. High quality, commercially available DC-to-DC converters are available from Brusa, Curtis, Sevcon, Todd, Vicor and others.

A quality DC-to-DC converter will be both compact and efficient at converting high voltage to low auxiliary voltage. Some lower quality units waste a significant fraction of electrical power as heat. Courtesy EVSource.com.

The Zivan line of DC-to-DC converters is popular with EV enthusiasts. Shown here is a 100–200 volt input unit that can provide a continuous supply of electricity at 14 volts and 60 amps. Courtesy KTA Services.

A Curtis 1400-series DC-to-DC converter. Courtesy KTA Services.

Power Brakes

All modern cars and trucks use hydraulic brakes to stop. In the simplest hydraulic brake system, a small piston attached to the brake pedal moves inside a master cylinder that is filled with hydraulic fluid. When the brake pedal is depressed, the corresponding movement of the piston in the master cylinder forces fluid to displace out the end of the cylinder and into a small diameter tube. This tube is branched off into four separate lines that are routed to the four wheels of the vehicle. At the end of each line is a small slave cylinder. A small piston in each slave cylinder moves in response to the fluid that is forced through the lines by the action of the master cylinder piston. The movement of these slave cylinder pistons is used to operate the brakes, which can be either drum or disk. In a drum brake, the movement of the slave cylinder piston causes a set of shoe pads to press radially outward against a rotating metallic drum that is attached to the wheel, thereby stopping the car by way of friction between the shoe and the drum. Disk brakes operate on a similar friction principle but rely on flat caliper pads clamping onto a rotating disk.

In reality, modern cars and trucks utilize brake systems much more sophisticated than the simple example described above. These modern brakes include separate (redundant) hydraulic circuits, anti-lock control computers, and electronic warning sensors. At their core, however, they all rely on the same basic hydraulic-mechanical principle, which begins with the physical movement of the brake pedal by way of the driver's foot.

Brake Boosters—As vehicles have grown larger and heavier throughout the years, the need for larger and stronger brakes has also increased. Large diameter, multi-piston all-wheel disk brakes have become the norm on today's two-ton cars and trucks. To help the driver physically apply the force on the pedal necessary to activate such large brakes, manufacturers have incorporated power assist devices, or *brake boosters*.

A brake booster is nothing more than a large diameter canister that resides between the brake pedal and the master cylinder. A diaphragm inside the canister separates its internal volume into

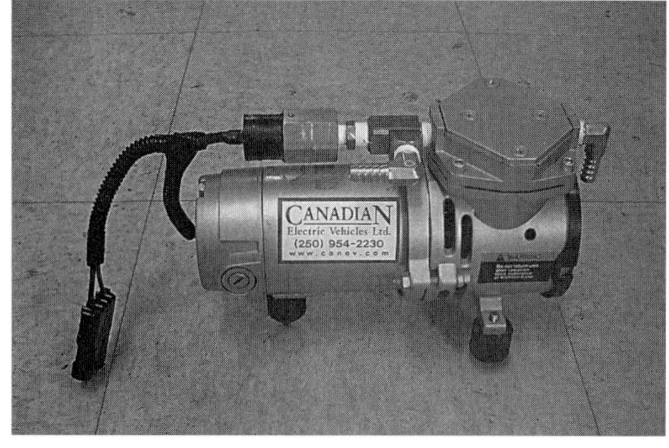

Modern cars and trucks are fitted with power brakes that are actuated by way of vacuum from the ICE intake manifold. Because electric vehicles do not have this ready-made supply of vacuum, an alternate means must be provided. A vacuum pump, like this one sold by Canadian Electric Vehicles, is more than capable of providing a vacuum source for a large power brake system. Courtesy Canadian Electric Vehicles.

While it is possible to cobble together everything needed to supply vacuum to a power brake system, it's much easier to simply buy a complete kit. Shown here is an aftermarket electric brake booster kit. Included are a vacuum pump, reservoir tank, automatic shut-off switch, power relay, fuse, and vacuum tubing. Courtesy EVSource.com.

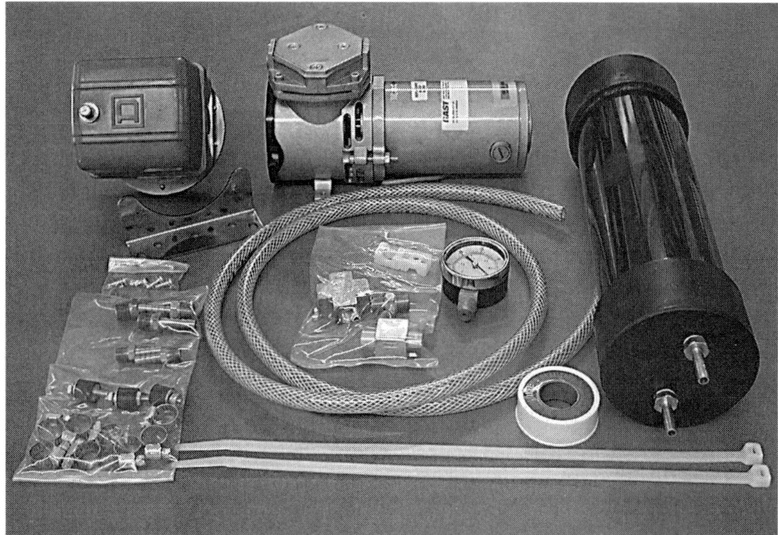

Another high-quality vacuum supply kit. Courtesy KTA Services.

halves. Both sides of the diaphragm are attached to a vacuum source, which on ICE-powered vehicles is usually the intake manifold. When the brake pedal is depressed, a small valve is cracked open which allows normal atmospheric air to enter the pedal-side volume inside the canister, but not the other side. Because of the resulting pressure differential, a force is created on the diaphragm that helps push the piston in the master cylinder, thereby providing the stopping power. When the driver's foot is removed from the pedal, the valve closes and vacuum is restored on both sides of the diaphragm again. Simple and effective.

Vacuum Pump—If the vehicle that you're converting to electric power has power brakes that you want to retain, you will need to supply a source of vacuum to operate the booster. Because there is no intake manifold that has an existing vacuum source that can be tapped into—because the internal combustion engine is removed—an alternate means of providing vacuum must be provided. Enter the vacuum pump.

A vacuum pump is nothing more than a device that creates a vacuum when powered by a 12-volt electric motor. When the pump is switched on, a partial vacuum is created at its entrance port. You can simply connect a small diameter rubber hose from this port to the brake booster and have power brakes as a result. Some vacuum pumps incorporate a small vacuum reservoir tank that provides a reserve source of vacuum. A small sensor and internal switch in the vacuum pump turns the pump off when a high enough vacuum level in the reservoir tank is reached. This keeps the motor from having to run continuously when not needed. As soon as the vacuum level drops below a certain threshold, the motor is automatically switched back on and the reservoir is recharged with vacuum.

There are a number of sources to consider when shopping for an electric vacuum pump. Do-it-yourself-type individuals are often able to find suitable used pumps in junkyards. Diesel powered vehicles, for instance, do not normally create vacuum in their intake manifolds, so they are a good source to consider when looking for pumps. If you want to buy new, most EV supply houses will carry at least one brand of quality pump, such as a Gast or Sensidyne unit. In addition, some suppliers manufacture their own in-house pumps.

Power Steering

A large majority of cars and trucks suitable for conversion to electric power come equipped from the factory with power steering. Modern cars are heavy, and at low speeds, the effort to physically

While not strictly needed for operation, a vacuum reservoir like the one shown here adds a level of sophistication and efficiency to a power brake system. The reservoir holds enough vacuum to keep the pump from frequently cycling on and off. Note the built-in vacuum pressure gauge that helps the operator detect leaks and ensures the system is providing sufficient vacuum. Courtesy EVSource.com.

A vacuum switch is used to detect when the vacuum pump and reservoir have developed sufficient vacuum. At this point, the switch turns off the pump. Courtesy EVSource.com.

A brake vacuum pump system installed in an electric vehicle. Courtesy Electric Car Company of Utah.

turn the steering wheel by hand can be significant. For drivers who aren't built like Arnold Schwarzenegger, the required steering force would be too heavy for normal everyday use, especially at low speed, which is when turning effort is the greatest in a car. The obvious solution for most manufacturers is to tap into the ICE to provide power assist to the driver.

A typical steering power-assist assembly on an ICE-powered vehicle includes a hydraulic pump powered by way of a drive belt from the engine, in much the same way that the alternator, fan, and air-conditioning are powered. When in operation, the steering-assist pump creates a flow of pressurized fluid that is directed to the steering box (or rack, depending upon the type of steering system) by way of high-pressure hoses. The flow of this fluid is controlled by valves operated by the motion of the steering wheel. When the driver turns the steering wheel to the right for example, a valve opens that sends pressurized fluid to an actuator that helps push the wheels to the right. When the driver stops applying a torque to the steering wheel, the flow of power steering fluid stops.

While this is a very reliable and effective means of assisting the driver in turning the wheels, there is a downside. Power-assist steering pumps require a relatively large amount of horsepower to function properly, especially at zero and low vehicle speeds. This power has to come from somewhere. On an ICE-powered vehicle, the power is supplied by the idling engine, which normally has excess horsepower to spare. You might think that a similar arrangement could be undertaken on an electric vehicle: why couldn't you simply build a bracket off the side of the electric motor to hold the OEM pump, and then drive the pump with a belt attached to the output shaft of the motor? Unfortunately, an electric vehicle, by its intrinsic nature, does not "idle". In other words, just when you need the maximum steering assist (at zero and low speeds) a belt-drive system wouldn't be functioning.

Okay, that's the bad news. The good news is that EV builders before you have solved this problem. The secret is to implement a standalone electric power steering assist unit, like the ones that Toyota uses on their late model MR2s and Spyders. These units are compact and come with an integral fluid tank, motor and pump, and controller unit. Be warned, however, that these assemblies are not cheap, plus you will have to have custom hoses made that connect the pump to the steering system, and you will need to fabricate a mounting bracket for the unit under the hood or in the trunk. Some builders have also built small electronic circuits that turn the pump unit off when the vehicle speeds rise and/or steering assist isn't required (e.g., when driving in straight lines) as a means to save electrical energy.

If you want power steering on your EV, you might consider using a late model Toyota MR2 Spyder steering pump. Traditional OEM-type steering pumps use power from the ICE to function, and therefore are a poor choice for an EV. The Toyota unit shown here operates with a built-in electric motor. The unit also includes a built-in reservoir for the steering fluid. Courtesy Bob Simpson.

The simplest way to address power steering in an EV is to convert a car that doesn't require it. A lightweight sports car or small vehicle equipped with a good manual steering system shouldn't need any upgrades or modifications when converted to electric power.

Power Windows, Seats, Locks, Stereo

Some electric vehicle builders strive for maximum mileage and efficiency. These hardcore types realize that every extra ounce their car carries and every onboard device and mechanism that uses electricity on their vehicle comes at the expense of range and performance. For these folks, luxury items like power windows, mirrors, locks, and seats are anathema to the reasons they drive EVs. Even stereos and defrosters are shunned. For these enthusiasts the three most important properties of their EVs are efficiency, efficiency, and efficiency.

For the rest of the population who drive EVs, however, the word *luxury* should be replaced with a question: "Is this item required or just desired?" The choice of which power equipment and accessories you decide to retain (or install) on your EV is entirely yours. No two owners are alike, and neither are their needs in a vehicle. For some people, power windows are a frivolous item, while other owners might consider them required. Take your time and determine whether you "need" a particular feature or accessory, or whether you simply "want" it prior to implementing it.

Similarly, where you own and operate your electric vehicle will influence whether you need to incorporate cabin climate control in your conversion plans. Having cabin heat in Los Angeles is probably not as critical as it is in Minneapolis. The same is true for air-conditioning, which can be absolutely essential in Phoenix, Houston, and Tallahassee, but not so important if you live in San Francisco, Des Moines, or Maine. Heat and cooling both require power to operate, and that power has to come from the onboard battery bank during normal operations. Said another way, if you want climate control—or any other "luxury" accessory—in your EV project, you will have to compromise range and/or conversion costs.

If your EV requires power steering, the simplest way to provide pressurized hydraulic fluid is a pump kit, like the one shown here. Included in this particular kit are a 12-volt hydraulic pump, reservoir, high and low pressure hose, circuit breaker and switches, and brackets to mount the unit. Courtesy EVSource.com.

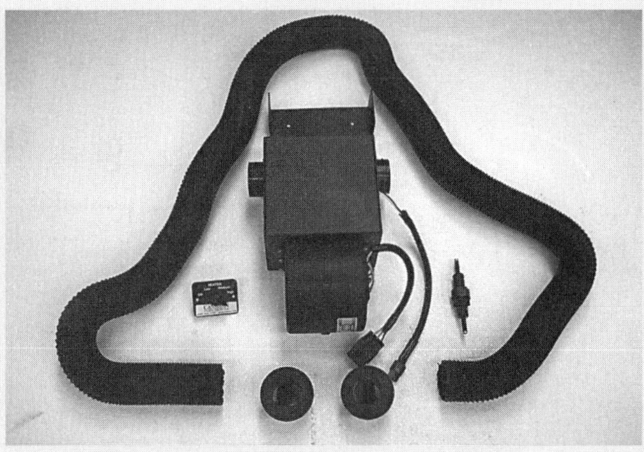

Cabin heat can be provided by way of an electric heating kit. Courtesy Canadian Electric Vehicles.

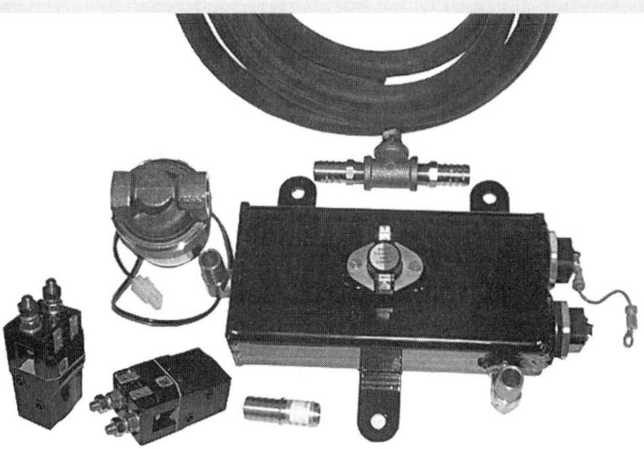

An in-line kit that heats water-glycol and circulates the warmed fluid through a vehicle's OEM heater unit. Courtesy Electric Car Company of Utah.

A European MES-DEA RM3 4 kW fluid heater used for cabin heat in an electric BMW project. The heater is capable of transferring 4 kW of heat into the water that is then pumped through the stock BMW under-dash heater core. The owner reports that there is little perceptible difference between the ICE-powered vehicle heat and defrost and the new electric heating unit. Says the owner, "The only real difference is the electric unit delivers heat faster than the OEM unit, probably because there is less thermal mass to heat up." Courtesy Bob Simpson.

In place of fluid-based systems, it is possible to convert your OEM heater unit to electrical power. A ceramic heating unit, like the 1500 watt unit shown here is installed in place of the original liquid-to-air heating core. Courtesy EVSource.com.

Air-conditioning in an electric vehicle can be accomplished by way of an electric A/C system. Shown here is a Masterflux Sierra compressor unit. Courtesy Masterflux/Tecumseh.

A control board for operating an electrically operated air-conditioning system. Courtesy Masterflux/Tecumseh.

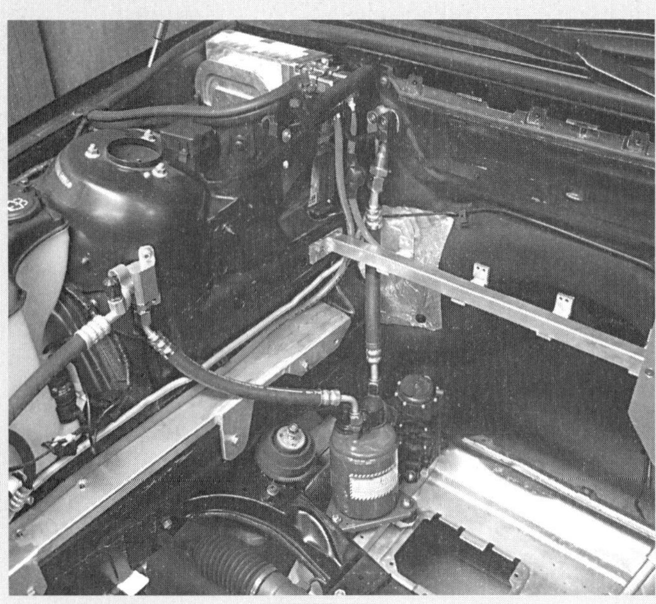

A Masterflux Sierra compressor installed in an EV. Courtesy Bob Simpson.

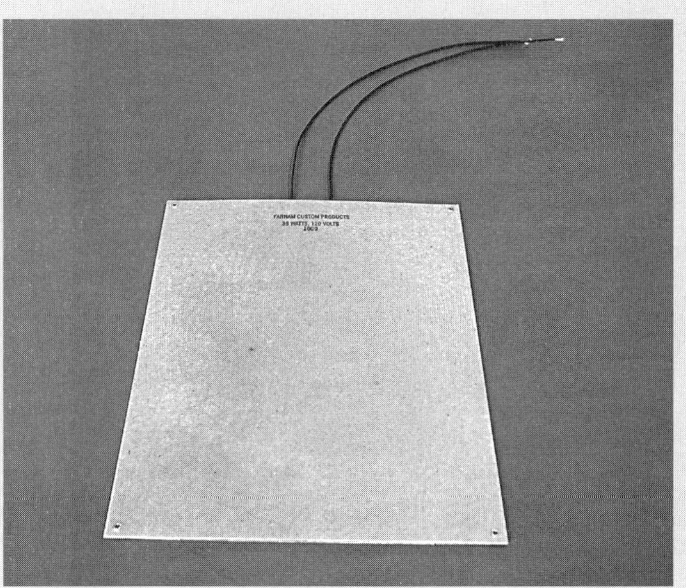

If you live in a cold climate, you might consider placing battery heating pads underneath your EV batteries. Shown here is a Farnam battery heating pad that operates on 120 volts input, and provides 35 Watts of heat. Courtesy KTA Services.

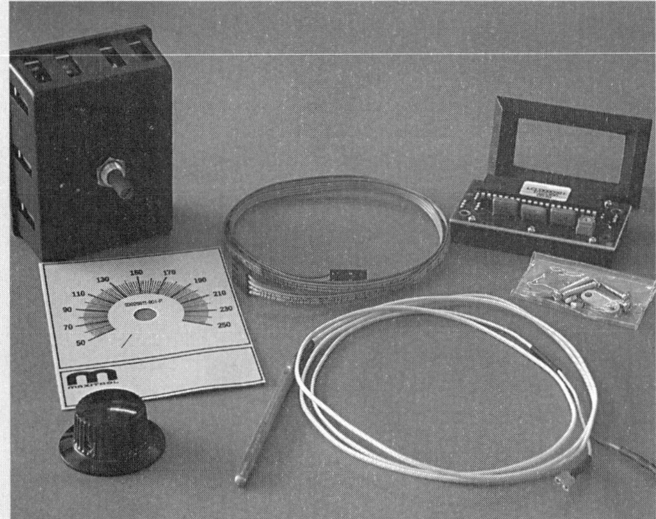

Shown here is a controller for a battery heating element. The unit comes with the electronic controller and a temperature probe. Courtesy KTA Services.

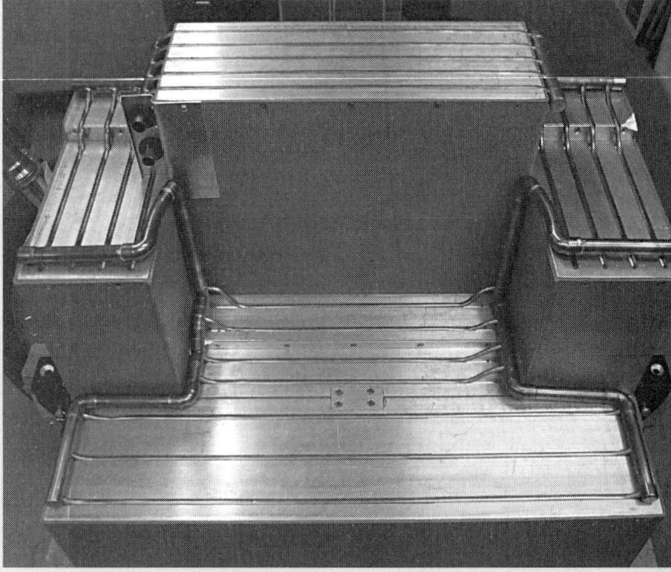

Depending on your climate, auxiliary heating or cooling may be required for your battery bank. The owner of this BMW required both for his vehicle. His solution was to build a water-cooled battery box. This photograph shows an underside view of the box that has an array of copper pipes that provide thermal conduction to the aluminum box. Courtesy Bob Simpson.

Chapter 10
Instrumentation

The two most basic gauges an EV requires are a voltmeter and an ammeter. Gauges allow you to monitor the performance of your vehicle. The primary requirements of gauges are reliability, accuracy, and readability. Quality should be a key consideration when looking at different makes and models that are available. Spending a little more up front on high quality, automotive-grade equipment is important. The extra money is often more than made up in the long run by ensuring accurate measurements. Being able to spot problems when they just begin—that is, before they turn into major issues—is vitally important to the long-term reliable operation of an EV. Courtesy Electric Car Company of Utah.

"If you put a meter on your dashboard that showed you how much it was costing you per mile [to drive an internal combustion engine-powered vehicle], I think most people would be shocked." —Sean Comey

When you remove the internal combustion engine and all of its supporting equipment from your EV conversion vehicle, you also do away with the need for many of the original dashboard instruments, gauges, and under hood sensors. The fuel gauge, water temperature, oil pressure, and other ICE-related instruments are no longer needed. Instead, EV-specific gauges are required. But which instruments and gauges should you consider installing in your EV project? While there is a lot of overlap between them, EV gauges fall into three general categories: state of charge, health, and performance.

Gauges that fall into the first category (state of charge) are in a sense the equivalent to a "fuel gauge" for an EV. A state of charge gauge is the means by which you can determine how much energy is currently stored in the battery bank, which allows you to predict how much range you have remaining before the vehicle requires a recharge. Voltmeters are the main types of state of charge gauges used in EVs, but there are other types available, too.

The second major category (health) includes the gauges and instruments used to monitor the instantaneous condition and status of the electrical system, thereby warning the operator when things begin to go amiss. Ammeters, voltmeters, tachometers, temperature sensors, and the like fall into this category that provides a snapshot view of the instantaneous operating state of the EV components.

The third group of gauges (performance) allows the operator a chance to monitor—and therefore affect—the performance and/or efficiency of the EV while it is underway. Ammeters fall into this category, as do specialty equipment such as E-meters and so-called "smart" gauges that measure instantaneous amp-hour usage and the like.

In this chapter, we will look at some of the different types

> **Selecting Gauges**
>
> When shopping for EV gauges, look for units that have large, readable displays than can be viewed when subjected to sunshine. Also, look for units that have backlighting so they can be read at night. Also, don't settle for non-automotive grade instruments. It's important to have a reliable, high-quality set of gauges designed for the harsh, vibration-prone environment of an automobile driven on poor roadways.

A Westberg 50–150 volt voltmeter. Simple, but very effective for keeping tabs on the voltage of a battery bank. Courtesy KTA Services.

of gauges and instruments that can be used in an EV, starting with the most basic instrument of all, the voltmeter.

Voltmeters

The most basic measurement you will need to make in your EV is the amount of voltage contained in the battery bank. The tool used for this is called a voltmeter, which is a device that measures the electrical potential difference between two points in an electrical circuit.

Voltmeters come in a variety of different types, from simple analog dial types that move a pointer across a scale in proportion to the circuit voltage, to sophisticated digital units that have auto-ranging abilities and display the voltage level on a numerical display. Generally speaking, digital voltmeters can provide somewhat higher accuracy than an analog type (1% vs. 2%, respectively). Digital voltmeters can also operate over a wider range of system voltages. Analog gauges normally must be selected based on the nominal operating voltage that is to be measured; i.e., a gauge that is appropriate for use in a high-performance 196 volt EV application, for instance, is probably not appropriate for use in an

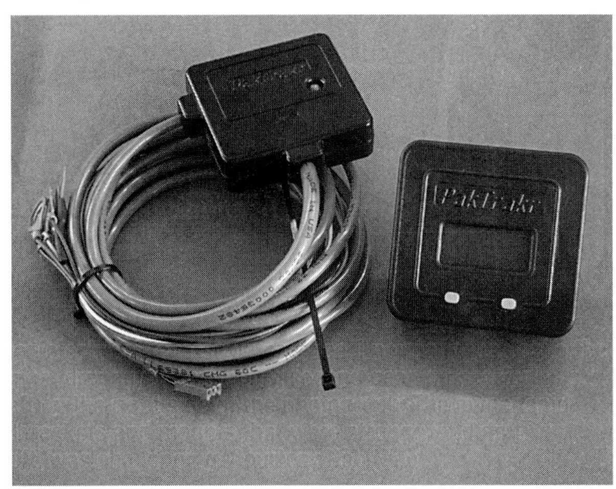

A Pak Trakr digital meter displays the battery pack state of charge and voltage, as well as individual battery voltages. Courtesy KTA Services.

> **Gauge Features**
>
> Features to look for in gauges include shielding against dust, moisture, shock, and vibrations. You should also look for units that have guaranteed accuracies of 2% or less.

economy-oriented 72 volt vehicle. Analog gauges, however, are often less expensive than digital gauges and are more widely available from a variety of manufacturers and in different physical sizes and levels of quality.

Besides measuring the voltage of the overall battery bank, voltmeters can be used to measure individual battery voltages, as well as voltage levels at various discrete locations in the EV, such as across the terminals on the motor. Some specialty voltmeters are known as "pack meters." These units can toggle between the overall system voltage and individual battery measurements with the push of a button.

When shopping for a voltmeter, look for a unit that can measure from zero volts up to approximately 10% higher than your expected nominal pack voltage. For example, if your nominal system voltage is 120 volts, select a unit that can measure up to 132 volts or so. This extra amount is needed because freshly charged batteries typically carry a surface charge of roughly ten percent over the nominal voltage. This charge disappears, or burns off, with the first few miles of operation and therefore shouldn't be included in normal range calculations.

After you've driven your EV for a few times, you will begin to figure out what voltage levels constitute "full" and "empty" for your application. Some operators place small pen marks or pieces of tape on their analog gauges to provide a quick reference.

A voltmeter used to measure overall battery bank voltage needs to be connected at the battery pack's most positive and negative terminals. This does not, however, mean that the voltmeters should be connected directly to the battery terminals. Doing so often results in corrosion attacking the voltmeter connectors and wiring. A better place to attach the leads from the voltmeter is on the main contactor (on the positive side of the system) and to an intermediate, large terminal block on the negative side. This is particularly true on systems that have a divided battery pack, with half of the batteries located at the front of the car and the other half at the rear.

State of Charge Gauges

Most state of charge (SOC) gauges used in EVs are based on the principle of measuring the battery bank's nominal voltage level. There are, however, other means of determining the amount of "fuel" remaining the pack. The four principle means typically used are: 1) voltage; 2) chemical, in which the specific gravity or pH of the batteries is monitored and then converted to an equivalent energy capacity; 3) current integration, in which the current flow from the battery bank is continuously measured and summed; i.e., the amount of energy removed from the battery bank is monitored; and 4) pressure, in which the internal pressure of some types of specialty batteries (such as NiMH) is measured and related to the effective remaining charge left in the battery.

Often called a battery "fuel gauge," a battery state of charge (SOC) meter is a simple way to tell how much useful charge is left in a battery bank. Courtesy KTA Services.

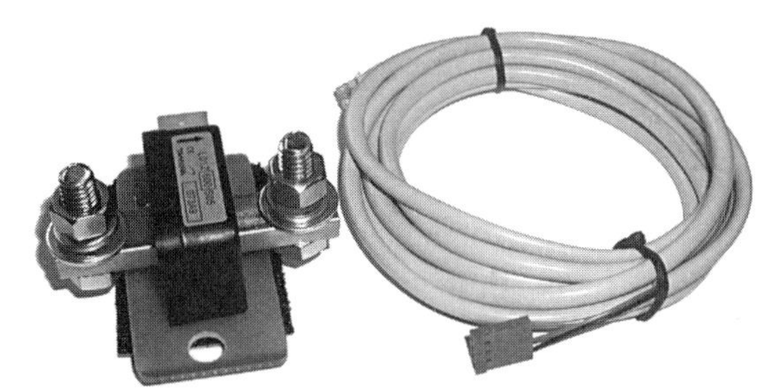

The Pak Trakr can be used to measure current with this amperage sensing kit. Courtesy EVSource.com.

State of Charge—A specialized type of voltmeter often used in electric vehicles is called a SOC, or state of charge gauge. Equivalent to a fuel gauge in an ICE-powered vehicle, a SOC provides a measurement of the remaining charge contained in the battery bank. In fact, many EV suppliers market their state of charge gauges as "battery fuel gauges."

The main difference between a conventional voltmeter and a SOC is that the latter does not necessarily display voltage as its output. Instead, a SOC may simply have a display that shows "Full" and "Empty," with intermediate gradations that represent percentages that the battery pack is depleted. Other types of SOCs have an array of light emitting diodes (LEDs) that illuminate in proportion to the amount of voltage the battery pack has remaining. In other words, while the parameter being measured is in fact voltage, the displayed output is in terms of usable battery energy left remaining.

State of charge gauges must be selected and/or calibrated to match the specific battery bank system voltage employed in the vehicle. It should also be matched to the allowed battery depletion levels. The depth of discharge of a battery affects its overall lifespan. If the battery is to be routinely drained to, say, 50% DOD during normal operations, the corresponding "empty" point on a SOC gauge will be different than in a vehicle where the normal DOD is only 80%.

Ammeters & Shunts

Besides voltage, the other primary measurement needed on an EV is current. Specifically, measuring the electric current flow to (and through) the motor can help the operator track and quantify energy usage, as well as diagnose small potential problems before they grow to large real ones. The means by which current is typically measured is via an ammeter, or ammeter.

Unlike voltage measurements, which are performed via a voltmeter attached in parallel to the electric circuit, direct current measurements require the sensor to be wired in series with the circuit. For small low-current circuits, this is in fact what is done; the circuit is cut and an ammeter is inserted directly in-line. The ammeter itself measures the rate at which the electrical charge moves through the circuit (via the flow of electrons) and displays this value on either an analog or digital gauge.

Unfortunately, this type of direct in-line measurement of current is only practical in small, relatively low-current circuits. When large electrical currents (like those that pass through the motor of a mid-sized electric vehicle in operation) flow in a circuit, a potentially dangerous situation would exist if we located an ammeter inside the cabin in the dashboard. Large currents passing into and

The chief means of measuring electrical current flowing to and from the battery bank is an ammeter. Shown here is a Westberg 0-1000 amp unit. Courtesy KTA Services.

A shunt wired in series with the main power feed is required for most high-amperage ammeters. Shown here is a Deltec 250-50 unit. Courtesy KTA Services.

through the cabin in direct proximity of human operators could result in a fatal situation if the line were to be severed, shorted out, and/or the ammeter were to fail. When laying out EV wiring, all high power wires and cables should be routed in short direct paths from the battery bank to the controller and on to the motor, and should be kept out of areas that the driver and passengers can accidently access.

So how do we measure current with a gauge mounted in the dashboard if we can't bring high current, high power into the cabin? The answer is called an *ammeter shunt*, which is nothing more than a very low resistance, very accurate resistor placed in series in the circuit. A voltmeter can then be connected in parallel to the shunt (one lead on one side of the shunt, and the other lead on the other side). The low resistance of the resistor means that virtually all the current in the circuit continues to flow through the resistor, as it is the path of least resistance. The voltmeter then is able to measure the small voltage drop across the shunt. Because the resistance of the shunt is accurately known, the measurement of voltage is correspondingly accurate.

But this is a measurement of voltage, not current. How do we get a reading of amps from a voltmeter? From basic electricity theory, we know that Ohms Law says that Voltage is equal to Resistance multiplied by Current, or V=IR. We can algebraically rearrange this formula to solve for Current; i.e., Current equals Voltage divided by Resistance, or I=V/R. Since we know the resistance of the shunt, and we can measure the voltage drop across the shunt, we can plug those values into the equation and solve for the current that is flowing through the shunt. And since the shunt is in series with the rest of the circuit—and therefore the amount of current flowing through the shunt is the same as that flowing through the circuit, we have a means of measuring current.

Now, to make things even simpler, if we have a matched voltmeter and shunt, we can simply re-label the dial face on the voltmeter to read out in units of amps and then calibrate the system. In other words, the voltmeter may in fact be measuring voltage, but the display that the operator sees is in units of amps.

Other Gauges

There are a number of other gauges that can be considered when planning an EV conversion project. For instance, most EVs are equipped with a 12-volt auxiliary, or accessory circuit used to operate lights, turn signals, horn, radio and other "normal" automotive electrical equipment. To monitor this accessory circuit, a 0–15 volt voltmeter is often used. On cars fitted with a DC-to-DC converter, this type of gauge may not be a necessity, but on vehicles simply equipped with a large 12-volt deep-cycle battery to run the auxiliary equipment, the gauge is probably a required device, especially if any of the 12-volt circuits are used to, say, power control circuits or other critical operations equipment.

Similarly, another type of gauge that some EV owners sometimes install is called an amp-hour meter (an amp-hour is a unit of electrical charge, with one amp-hour being equal to the electric charge transferred by a steady current of one ampere for one hour). An amp-hour meter is much like an ammeter but instead of displaying the instantaneous amount of current flowing through a circuit, it displays the cumulative number of amp-hours that passes through the circuit. In a sense, an amp-hour meter is a type of fuel usage gauge, counting up (or down) the amount of energy that passed through the circuit at the location that is measured.

Amp-hour meters come in different makes and models and with different levels of sophistication. Most are digital and often come with multiple output features, such as state of charge, amp-hour usage, and estimates for remaining amp-hours and operating time left on the battery. Like an ammeter, an amp-hour meter requires the use of a shunt wired in series in the circuit.

Motor tachometers are also sometimes installed to help monitor shaft speed, ensure that the motor does not over rev, and determine the correct shift points. In addition, temperature sensors are also sometimes used to monitor critical components, such as controllers and motor housings.

Sometimes the simple approach is the best way to instrument an EV. This electric roadster features just three basic gauges: an ammeter, a state-of-charge meter, and a speedometer. Courtesy eRoadsters.com.

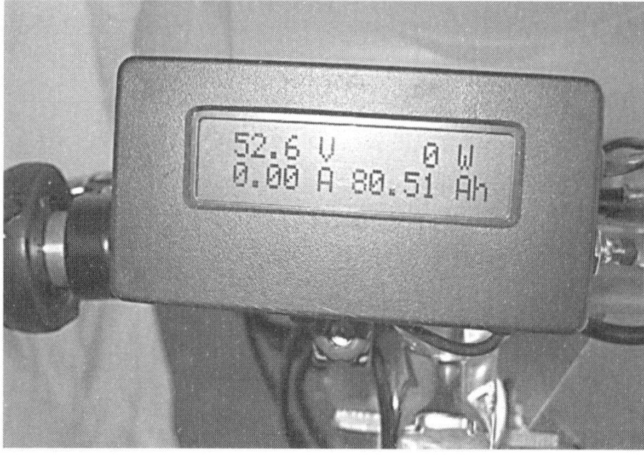

The Cycle Analysts is a digital dashboard and battery monitor. Originally intended for use on electric bicycles (as shown here), the Cycle Analyst has gained favor in other EV applications, including scooters, motorcycles, and even cars and trucks. The unit displays instantaneous voltage, Watts, and amps. It also provides cumulative readings for amp-hours and watt-hours used; speed, distance, time, peak current drawn; voltage sag; total battery cycles; and amp-hours. It can even track regenerative braking energy and total watt-hours/distance traveled.

Shown here is a Blue Window digital gauge that can monitor and display battery voltage, current motor speed, and battery capacity. The display is simultaneously in both analog and digital format, with the output displayable either on a dedicated LCD monitor, or, as shown here, through an in-dash DVD player. Courtesy KTA Services.

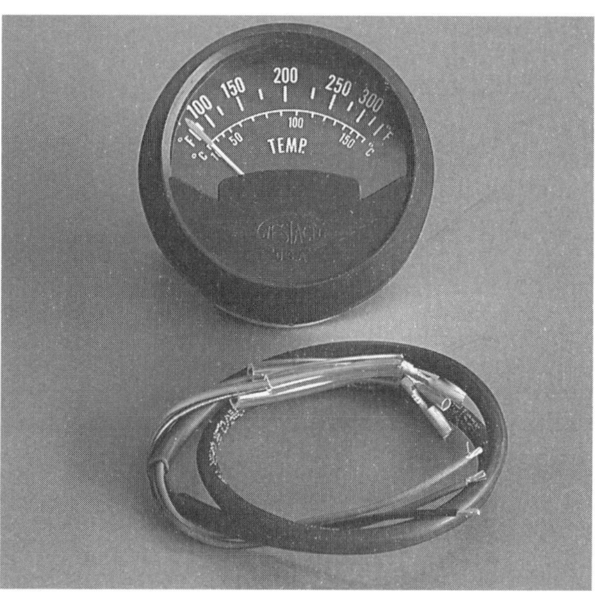

A number of EV components can get warm when operating normally. Overheating, however, is not desirable. Shown here is a Westberg temperature sensing unit that can be used to monitor the operation of key equipment such as controllers, motors, and chargers. Courtesy KTA Services.

Motor speed can be displayed by way of a tachometer, such as the Westberg unit shown here. Courtesy KTA Services.

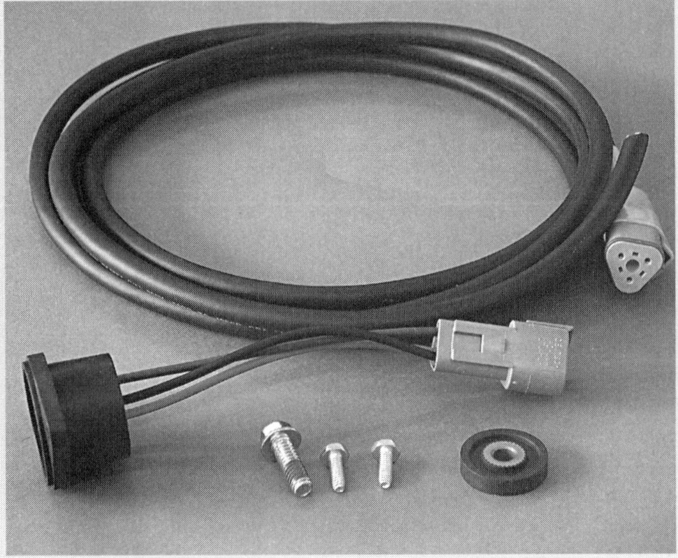

A motor shaft speed sensor kit. Courtesy KTA Services.

A Zolox motor shaft speed sensor and mount. This unit affixes to the end of an electrical motor and provides a signal to a display unit. Courtesy EVSource.com.

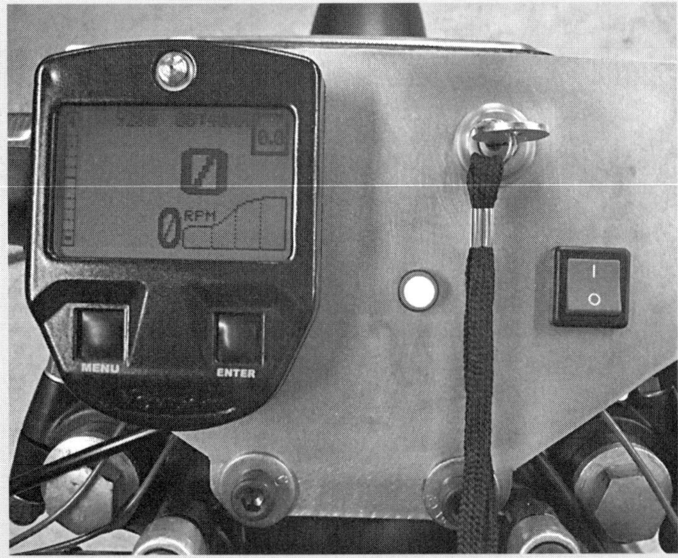

A Veypor multi-function motorcycle gauge that includes a digital and analog speedometer, odometer with trip distance, quarter mile timing, 0-60 mph timing, efficiency, and other outputs. Courtesy Tony Helmholdt.

PROJECT VEHICLES

Now that we've discussed all of the individual parts, components, and subsystems of an electric vehicle, it's time to see how all the pieces come together in some real-world examples. The following nine project vehicles represent a wide range of complexity and cost. Each owner had a different set of goals in mind—and budget to work within—and consequently arrived at his own unique solution to electric transportation. While the vehicles on the following pages may appear totally different from each other, all had two things in common: a well-thought-out plan and strong attention to detail. Study what these nine enthusiasts have done with their vehicles; whether you are new to electric vehicles or are an old hand, there are valuable lessons to be learned here.

Project 1:

Mountain Bike

A brand new 2009 HASA Team EXD full-suspension mountain bike converted to electric power. "When planning what parts to use in my conversion," says the owner, "I needed to first establish my requirements. This was done by working out round-trip commute distances I needed each day and the allowable amount of time I had to travel those distances. This resulted in a max range requirement of 50km, and a speed of 40km/h. From my research on the internet, similar bikes were consuming approximately 15Wh/km. I used this number to size my battery pack. 15Wh/km x 50km = 750Wh. Knowing this, I found a battery with a capacity of 768Wh and was set." Courtesy James Parsons.

Of all the possible project vehicles that can be converted to electric power, bicycles are perhaps the easiest for a beginner to undertake. They also offer arguably the best possible bang-for-the-buck return for someone looking for an inexpensive, but reliable (and fun) means of transportation. What could be simpler than bolting a motor, battery, and controller to a bicycle and zipping off down the road? Or more to the point, what could be more fun? Charge the battery, get on, and then twist the throttle—away you go. The speed and range of modern EV bicycle conversions are more than adequate for most work and school commutes. Even better, electric bicycles can ride in marked bike lanes, drive down alleys and bicycle trails, and park anywhere a regular bike can. For these and other reasons, electric bikes are becoming more and more popular in both urban and suburban settings alike. Electric bikes simply make a lot of sense for a lot of people who need to get from point A to point B.

These were the thoughts that James Parsons of Melbourne, Australia had when he began contemplating the construction of an electric bicycle. James needed an inexpensive, but reliable means of commuting the 50 kilometers (30 miles) roundtrip to and from work every day. Electric bicycles from China were showing up in his hometown, but while intriguing, the majority of these bikes were either expensive, poorly made, or both. If he wanted high-quality at an affordable price, James realized that he needed to convert his own EV bike.

The project began with one week of intensive online research, reading about other bicycle conversion projects and pricing kits and components from different suppliers. Then all the parts were ordered, including the bicycle itself, which was purchased new specifically for the conversion. "In less than two weeks," James says, "I went from researching the project to having a new bike and boxes of EV parts sitting on my living room floor. Three weeks later, I had a working electric vehicle. When planning this project, I took my inspiration from many other electric bike conversion I found on the internet."

James explains that after the parts showed up, he broke the project down into three distinct phases: 1) attaching the heavy battery securely to the frame; 2) installing the rear wheel hub motor to the bike; and 3) wiring everything together. "I started by mounting the battery, which is a 48-volt $LiFePO_4$ unit, at the rear of the bike. The battery is big and heavy, so I needed to reinforce the rear rack to carry the load." James built a brace for the cantilevered rack using thick tubular steel, rubber padding, and U-bolts. The battery itself was mounted inside a steel military box that can be locked. "The most expensive part of the conversion was the battery," James says. He also added a modified motorcycle alarm that has a vibration sensor and a two-way communicator with remote keychain. "This is important. I like knowing if someone is messing with my bike when I'm in a store."

With the battery securely mounted, James turned his attention to the hub motor, which is a Crystalyte unit from China. To attach the hub motor to the wheel, James used heavy-duty 13-gauge spokes. This required a small enlargement of the hub motor spoke holes with an electric drill. That was the easy part, says James; the hard part came later when he tried to bolt the motor into the bike frame. "One of the most difficult aspects of the conversion was getting the wheel and motor to fit correctly between the rear dropouts on the frame. This was partly due to the fact that I was using a rear disk brake, which is not typical in these types of conversions. Mounting the wheel took a lot of trial-and-error, using different spacers and washers. It was a challenge, but I'm happy with how it turned out."

Finally, it was time to wire up the bike. A Crystalyte pedal-first type of controller was chosen because it was a good match for the 48-volt lithium battery. The controller also requires the user to start manually pedaling before it kicks in. This system eliminates the need for hall effect sensors on the motor, which is something James says has been an issue with some other Crystalyte motors. This feature also acts as a type of safety device, preventing power to the motor when the bike is at a standstill.

The charger that James used was the one included with the purchase of the battery. He also installed a Tyco 48-to-12-volt DC-DC converter to power accessories, such as the custom LED headlight and the bike alarm. A digital Cycle Analyst instrument was also procured and then mounted to the handlebars. "The Analyst shows everything from current draw to pack voltage to watt-hours consumed. It even shows the speed of the bike." A 500-amp automotive-type blade fuse was installed to safeguard against short circuits.

Since he's converted the bike to electric power, James has begun riding it to work on a daily basis. He estimates that by combining the EV bike with public transportation, he is able to achieve nearly 90% of his transportation needs without using his car. The bike can achieve 42 km/hour (26 mph) on flat ground and has little trouble with even the steepest of hills if pedal assist is employed. James has also routinely achieved 50 km distances on a single charge at full throttle. "The great thing about an electric bicycle is that it has potentially unlimited range," says James. "I just pedal more if I want to go even farther."

SPECIFICATIONS

Vehicle: 2009 HASA Team EXD full-suspension mountain bike
Original Engine: n/a
Motor: Crystalyte Model 407 brushless DC hub motor
Motor Adapter: n/a
Controller: Crystalyte 24-72V 20A pedal-first controller
Batteries: Ping-made 48V 16Ah lithium-ion phosphate ($LiFePO_4$), with integrated battery management system (including low voltage, high voltage, and current limits)
System Voltage: 48V
Charger: Ping-supplied 2 Amp charger
DC-to-DC System: Tyco 50W 48V to 12V
Instrumentation & Gauges: Cycle Analyst
Safety: Pedal-first type controller
Drivetrain Modifications: Custom disc brake adaptation to hub motor
Suspension Modifications: Readjusted suspension springs to carry heavier load
Body Modifications: Reinforced rear bike rack
Interior: n/a
Charging Time: 8–10 hours
Range: 50 km (30 miles)
Top Speed: 42 kph (26 mph)
Acknowledgements: "I'd like to thank Justin Wright for helping fit the hub motor to the dropouts, Stef Wright for helping choose the right bicycle and assisting with the brake setup, and Colin Moore for helping to reinforce the battery box. I'd also like to thank Zev from www.ebikes.ca for helping to choose the motor/speed controller combo, and Sealys Cycles in Frankston for lacing the hub motor to the 26" rim."

The motor chosen for the conversion was a model 407 brushless DC hub motor from the popular Chinese company Crystalyte. The 408 model was also considered, as it can be pushed well beyond its stated maximum power. In the end, however, the owner chose the 407 model instead because the unit has a slightly higher top speed than the 408. Note the use of the disk brake mounted to the left side of the motor. Courtesy James Parsons.

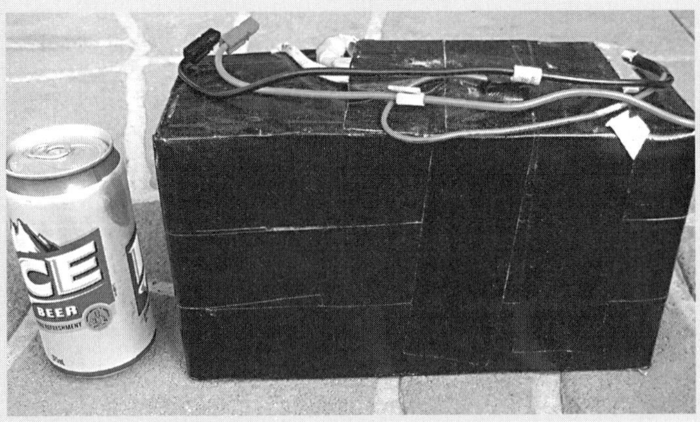

When asked what has impressed him the most about the bike, the owner does not hesitate in answering: "Definitely the battery. Lithium batteries, specifically the LiFePo4 batteries, are the best choice for these types of bike conversions. They hold a large amount of energy for their weight and are able to run strong until practically empty. Without this battery, I never would have been able to meet my range and speed requirements." Courtesy James Parsons.

The battery, controller, and DC-DC converter are mounted to the steel army surplus box, which in turn is attached to the rear rack behind the rider's seat. The owner built a small support brace underneath the rack to triangulate the bracket and provide enough strength to carry the heavy load of the battery and equipment. Courtesy James Parsons.

All instrumentation needs are provided by a digital Cycle Analyst attached to the handlebars. This unit displays everything from motor current draw, to battery voltage, to energy consumed, to the bike's current speed. Lightweight, simple to install, and easy to use, this type of display system is recommended for anyone considering an electric bike conversion. Courtesy James Parsons.

A DC-DC converter is used to supply 12-volt power to bike accessories, such as the custom headlight shown here. To build the headlight, the owner used three CREE XR-E light emitting diodes (LED) housed in an aluminum shell attached to the handlebars. Courtesy James Parsons.

Project 2:
Motorcycle

A very clean example of an electric motorcycle conversion project. The original bike was a new 2007 Lifan 200cc commuter bike imported from China. The owner wanted a non-polluting, economical commuter vehicle that someone could drive to work and school on a daily basis. Courtesy Lennon Rodgers.

In *Zen and the Art of Motorcycle Maintenance*, author Robert Pirsig wrote, "The place to improve the world is first in one's own heart and head and hands, and then work outward from there." In other words, if you want to clean up the world, start in your own backyard before asking your neighbors to clean up theirs.

Lennon Rodgers, the designer, builder, and owner of the novel electric motorcycle shown here is an adherent to this Think Globally, Act Locally philosophy. When he first considered building an EV, Lennon was living in Los Angeles and commuting to work by bicycle. He was also suffering from smog-induced asthma. Everywhere he looked when riding his bike, gasoline- and diesel-powered cars and trucks were spewing pollution into the air. One day he asked himself how he could help improve the air quality in LA, starting with a reduction of his own carbon footprint. The answer was an electric motorcycle.

"I vividly remember riding my bicycle and breathing tailpipe exhaust in Pasadena when the e-motorcycle idea hit me," says Lennon. "I have always been a motorcycle enthusiast and I knew that an electric conversion made a lot of sense. I decided to sell my gasoline-powered motorcycle and build an electric-powered replacement. I wanted to show that an electric motorcycle could be built inexpensively and be very reliable. I knew that EVs would be a great way to lower the pollution in the air in cities like Los Angeles."

Lennon started his project by undertaking a self-taught crash course in power electronics and EVs. He learned how to calculate key vehicle parameters and model the performance of different EV solutions. Lennon says, "After a few long online sessions, I had my head wrapped around the basic equations, which meant I was able to design the motorcycle on paper and with a computer spreadsheet program."

It took Lennon about a week to size a motor, choose the number, type, and capacity of the batteries, decide on a chain gear ratio, and finalize the basic electronic components list. He then needed a motorcycle frame to bolt all these pieces onto.

While there were numerous used motorcycles available in the local paper and on the internet that could have served as perfectly good starting points for the project, Lennon chose to begin his conversion with a brand new motorcycle made by Lifan. "I purchased a new motorcycle from an online supplier because I wanted others to have the ability to build the exact same design as mine. They could order the same Lifan over the internet and build a clone of my bike."

Another benefit of choosing a new motorcycle as a platform for conversion is that the builder can focus solely on adapting the motorcycle to electric power and not have to worry about things like rebuilding brakes, replacing bearings, and fixing other parts to ensure the bike is roadworthy.

SPECIFICATIONS

Vehicle: 2007 LIFAN TMS motorcycle
Original Engine: 0.2-liter 1-cylinder gasoline internal combustion
Motor: E-TEK RT, Model number ME0709 DC motor
Motor Adapter: Custom owner-built
Controller: Kelly Controller: KD72301 with Regen; two Magura 0-5K Twist Grip throttles—one for acceleration, the other for regen/deceleration
Batteries: Six 12V B&B EVP44-12 sealed lead-acid batteries
System Voltage: 72V
Charger: Six individual Soneil 12V 3 amp constant chargers
DC-to-DC System: Kelly
Wiring: #4 welding wire for main power runs
Instrumentation & Gauges: Cycle Analyst CA-HC
Drivetrain Modifications: Custom 72-tooth rear sprocket; 12-tooth front sprocket
Suspension Modifications: None
Body Modifications: Gas tank modified for charging plug
Charging Time: 6 hours
Range: 25–30 miles
Top Speed: 45 mph
Acknowledgments: Rick Rodgers

Purchasing a new motorcycle would allow these steps to be bypassed from the start.

Lifan is a low-cost brand of commuter motorcycles that are made in China and sold online. Fortunately for Lennon, these motorcycles are distributed in Los Angeles, and he was able to drive to the warehouse and look the bikes over firsthand. He then ordered one delivered without an engine. Approximately on month later, Lennon had his motorcycle and all the major conversion components delivered to his house.

Cardboard mockups of the batteries, motor, and even the rear sprocket were used to decide on mounting points and packaging. "I decided to use six 12-volt batteries to power the motorcycle," says Lennon. "At first it seemed impossible to fit all that volume and mass on to the small steel Lifan frame, but I took my time and worked through the issues." Once everything was decided, mounts were fabricated and welded directly to the frame.

After he had tackled the basic packaging of the larger components, Lennon turned his attention to bench testing the electronics. The batteries, motor, controller, and other equipment such as the regenerative braking system were wired up and checked out prior to installation. Then it was time to transfer the equipment from the bench to the bike. Two weeks later, most of the equipment was mounted and the wiring was complete.

"Late one night I realized I was ready to test. I opened the shop doors, turned the key, and twisted the throttle. It worked! My first test drive made me feel like a real pioneer." Lennon then took the next few weeks to put the finishing touches on the bike, such as hard mounting switches, tidying up the wiring, and so on. Then it was time for his first serious trip. Lennon drove just under 30 miles by cruising around Los Angeles. "As I passed through some of the rougher neighborhoods of L.A., I was confident that my triple checking of everything, and my perfectionist soldering, crimping, and welding would allow me to arrive home safely."

At the time of this publication, Lennon has put over 1000 miles on the motorcycle. He is so pleased with the results of the conversion that he has made the plans for the bike available for free on his www.electricmotion.org website. "I wanted to build an EV, but at the same time I wanted to encourage others to also do so. I thought it was important to buy all the parts on the Internet and make the plans available for free. Electric vehicles like this one make a lot of sense for a lot people. I've shown it can be done, and now I hope many others do it, too." We have no doubt this wish will come true. We also suspect that Zen author Robert Pirsig would be proud of Lennon's creation. He might even want to take the bike for spin around his own neighborhood.

The motorcycle was purchased new and delivered without an engine. This eliminated any restoration or rebuilding to make the bike roadworthy prior to conversion. Note the cardboard mockup of the electric motor test fitted to the frame. Similar mockups for the batteries were used to help determine a method of mounting them to the frame. Courtesy Lennon Rodgers.

A cardboard mockup of the 72-tooth rear drive sprocket was also used early in the project. When selecting EV components, the owner says that the major trade-off he dealt with was between range and speed. In its current configuration, the bike can go roughly 30 miles before requiring a recharge and has a top speed of around 45 mph. The front 12-tooth sprocket can be swapped out for a larger unit if a higher top speed is required, but this comes at the price of lower range. Courtesy Lennon Rodgers.

The owner modifying the frame and grinding the welds smooth. A lot of thought was invested before the first cut was made. Says the owner, "It almost seemed impossible to fit all the volume and mass [of the batteries] on the frame." Taking his time and planning everything out carefully before acting was key. The owner also reports that having access to shop space at the local university was invaluable. Courtesy Lennon Rodgers.

In this photograph, the E-TEK RT motor is mounted securely to the frame, and the first pieces of the battery mounts are welded in place. The owner enrolled in a welding class at a local community college to learn how to join metal properly. Says the owner, "Many people are constrained by their lack of welding skills. I was no different, and I knew I needed to learn, so I did." Courtesy Lennon Rodgers.

A view showing the main brackets for mounting the six 12V sealed lead-acid batteries to the frame. All of the power electronics were bench tested before installation on the motorcycle. The owner says, "Overall, the most important aspect to EV building is to not take any shortcuts. Be very particular about every single connection, weld, and bolt." Courtesy Lennon Rodgers.

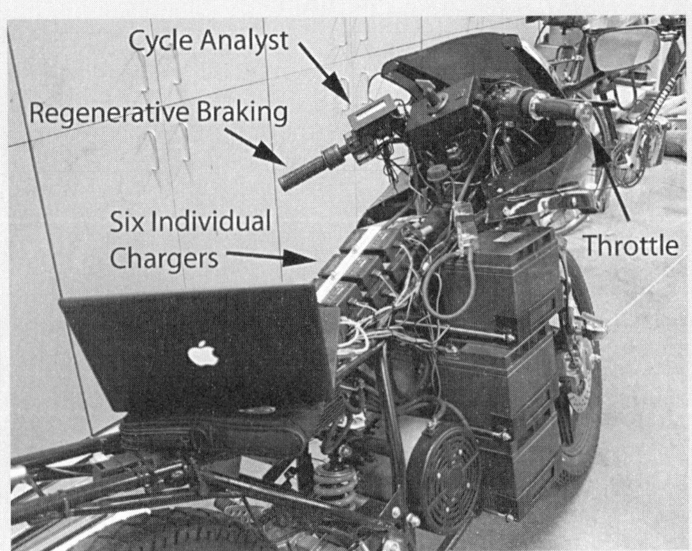

Six individual Soneil 12-volt chargers were employed to ensure proper recharging of each of the six batteries (i.e., no overcharging) Time to fully charge the system is approximately 6 hours. Also shown in this photograph is the regenerative braking switch and reverse "throttle" mounted to the left side of the handlebar. Twisting the grip forward varies the amount of regen possible. Mounted next to the regen control is a Cycle Analyst CA-HC unit that displays real-time energy usage, battery voltage, and electrical current draw. Courtesy Lennon Rodgers.

From start to finish, it took the owner approximately three months to convert the motorcycle. "Battery technology has changed since I built the motorcycle," says the owner. "If I were doing it again now, I would consider purchasing some of the low-cost lithium-ion batteries that are made for high-power applications." Note the plug on the top of the "gas tank" for recharging the batteries. Courtesy Lennon Rodgers.

The completed electric motorcycle weighs in at 480-lb, which is mostly due to the six heavy lead-acid batteries. Performance is very good, however, with brisk acceleration and excellent braking. After completing the project, the owner has made the construction plans available for free on his website www.electricmotion.org. "I want to encourage others to build low-cost EVs like this one. It's a lot of fun and not too hard to undertake. Riding the motorcycle is a blast, too." Courtesy Lennon Rodgers.

Project 3:

Bugeye Sprite

A very clean example of a 1962 Mk2 Austin Healey Sprite that has been fitted with an earlier model Mk1 "bugeye" bonnet (hood). The little car (dubbed "Little Go Green" by his wife) is equipped with an Advanced DC 8" motor and ten 12-volt Trojan lead-acid batteries. Range is a very respectable 40 miles, with a top speed of over 70 mph. Acceleration is also very good; the owner has more than once bested stock ICE-powered Sprites at the local drag strip. Courtesy Mark Hayes.

In 1958, the British Motor Corporation (BMC) rolled out a tiny two-seat roadster that they were planning to produce and sell. Named the "Sprite," the diminutive car had been designed by the Donald Healey Motor Company, and, to keep production costs down, utilized a number of components and parts borrowed from other cars that BMC and Healey built. The Sprite was intended to be an entry-level sports car that was fun to operate, looked good on the road, and was practical to own. With its mildly tuned version of the then-ubiquitous one-liter Austin A-Series four-cylinder engine, a simple but robust suspension and steering system, and an unadorned interior, the two-seat roadster was marketed as an inexpensive sports car that any "ordinary chap could keep in his bike shed."

Which brings us to just such a chap. Owner Mark Hayes of Boise, Idaho is an unassuming mechanical engineer who seems to have directly channeled Donald Healey's ghost when planning this novel 1962 MkII Sprite electric conversion project. As Mark puts it, "I wanted to build an electric car that was both fun and practical. Just like the original Sprite, I wanted to keep my car as simple as I could. I didn't want to reinvent the wheel. Wherever possible, I wanted to use proven components and off-the-shelf parts. Keeping the cost as low as possible was also very high on my priority list."

Mark originally received the Sprite from a friend, who had stripped the car down to sheet metal and then actually stored it in a proverbial shed. Unfortunately, stray fireworks set the storage shed on fire and destroyed everything on the car except the chassis and hood. Mark had done some repair work on a Datsun roadster for his friend, who traded the remains of the Sprite in payment for this help. Mark knew that he wanted to restore the car, but a busy life as an engineer and family man meant that the parts would have to be stored for a while. Unfortunately, "a while" turned into more than two and a half decades before Mark was able to restore the car. Says Mark, "I dreamed of restoring the Sprite ever since I got the parts, but I never was able to find enough time to commit to the project."

Mark goes on to say that he had also wanted to build an electric vehicle ever since he was a teenager growing up through the gas crises of the seventies. The magazines *Popular Science* and *Popular Mechanics* were filled with stories of electric cars and homemade conversions, and Mark never forgot the desire to build his own EV. Years later, when a window of opportunity to work on the Sprite finally arose in 2007, Mark at long last saw a chance to make this dream—plus the one where he restored the Sprite—both come true at the same time. He decided to resurrect the Sprite, but make it electric powered in the process.

Hardboard Masonite mockups of the batteries were used to help plan the conversion. On a tiny car like a Sprite, this planning step is invaluable, as it allows the builder to try various layouts and configurations without worrying about supporting the actual (heavy) batteries themselves. Courtesy Mark Hayes.

"After years of waiting, I finally began," says Mark. He started the ambitious process by choosing a motor and controller and selecting the batteries he wanted to use. Simple cardboard mock-ups of the batteries were constructed that helped Mark figure out where the bulky units might be located in the pint-sized car. "I originally wanted to use 16, or even 18, Trojan T145 6-volt batteries, but I quickly discovered that there simply wasn't enough room in the car. This led to my decision to use ten Trojan T-1275 12-volt batteries instead."

Once the basic components were selected and a budget developed, Mark ordered the major conversion parts. He also sourced all the regular restoration parts needed, such as a transmission, flywheel, suspension, and brake parts. The Sprite chassis was taken out of storage, cleaned up, and the damaged sheet metal from the fire was cut off and replaced. Then the real fun began.

"Once the parts started arriving, I had a few problems. For instance, the motor adapter plate had the wrong bolt pattern. Also the transmission pilot shaft interfered with the motor shaft." All of this necessitated some redesign and rework.

Once these individual problems were solved, Mark turned his attention to designing and building the motor and battery mounts. The actual motor was used to mock-up its mounting, but for the batteries Mark built a second set of mock-ups, this time from lightweight Masonite hardwood. Constructing lightweight battery "boxes" allowed a number of different packaging layouts to be tried. "I focused on making sure that the weight of the motor, transmission, and front battery box would be fully supported at the original motor mounting and transmission mounting points. I knew these hard locations were originally designed to carry the ICE loads, but I also reinforced them to increase their strength."

A tubular steel subframe was also constructed at the rear of the vehicle to support the weight of the middle and rear battery boxes. This frame was provided with attachment points for a roll bar, rear bumpers, and for possible future coil-over shocks. In addition, the subframe was designed to act as a conduit for much of the wiring. During this process, Mark cut and rebuilt much of the rear chassis sheet metal.

Once the batteries and motor were mounted, attention was turned to the mounting layout of the big Curtis controller, the DC-to-DC converter, and all the relays, contactors, high current shunt, and other equipment. "I bolted these components on to a thick aluminum plate that would act as a heat sink. Unfortunately, it took three tries to finally converge on the final configuration."

Mark then began some preliminary testing without fully finalizing the mounting of the components or finishing the restoration. "I started testing with the car still in a rough, unfinished stage. I roughed in the wiring for the drive and charging circuits so that I could debug those systems early. I called this my 'prototype' phase, and it really helped discover problems and allow me to address them in a straightforward manner."

The car was then towed to a large, empty parking lot and initial test drives were performed. This early testing led to a few changes to the battery configurations, as well as other modifications, such as changes to some of the suspension components.

"Once I had the final layout of all the parts and had the car running well, I stripped everything down and started the final restoring and finishing phase of the project." Mark spent a lot of time during this period laying out the final wiring for the drive and charging systems, as well as the 12-volt accessory system. A Painless Performance Products fuse center made this effort relatively easy to accomplish.

After the car was back together, each of the circuits was brought online one at a time. This made any final debugging very straightforward to troubleshoot. The final testing of the finished car also went very effortlessly, primarily because so much early testing was performed during the prototype phase. There was one small problem, however, that required attention: vibration.

After driving the car for a while, Mark found that the motor exhibited some vibration at certain motor speeds even though he had dynamically balanced the motor/flywheel/clutch assembly on the bench before installing it in the car. The bench balancing was done by powering the motor directly with a single 12-volt DC battery and adjusting weights on the flywheel according to a dynamic

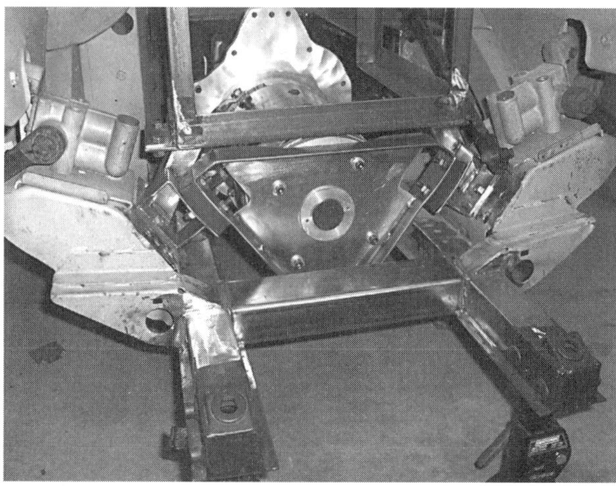

To fit the front batteries into the engine bay without modifying the bonnet (i.e., the hood), the owner remounted the transmission a few inches lower than its original location. The electric motor mounting plate was then carefully constructed to ensure that the weight of the motor and transmission at the front of car, along with the front battery box, would be fully supported on the original internal combustion engine mounting locations. These hard points were then reinforced with gussets that were welded in place. Courtesy Mark Hayes.

balancing program found on the internet. To correct the problem, Mark removed the front battery box from the car, pulled the motor/transmission assembly out of the car, removed the transmission, and reinstalled the motor assembly in the car. He then connected the remaining six batteries in the car to the controller. Using the motor controller and tachometer, he was able to find the motor speeds that displayed the most vibration and then dynamically balanced the motor assembly at those speeds. The result was a smooth quiet running motor at all speeds.

Mark now drives the Sprite to and from work daily and around town and has shown it in numerous car and roadster shows. The Sprite has also been featured in newspapers and on the local news. Mark says that every time he drives the car, at least one passerby waves or honks. The Sprite was so much fun to build and drive, says Mark, that he has started another conversion project, a 1967 Austin Mini Countryman. Evidently one small electric-powered British car isn't enough for this ordinary chap.

SPECIFICATIONS

Vehicle: 1962 Austin Healey Sprite MkII
Original Engine: 0.95-liter 4-cylinder gasoline internal combustion
Motor: Advanced DC 203-06-4001 8" DC motor
Motor Adapter: KTA / Electro Automotive adapter
Controller: Curtis 1231C
Batteries: Up to ten 12V flooded lead-acid Trojan T-1275 (currently using nine)
System Voltage: 108V
Charger: K&W #BC-20 onboard charger
DC-to-DC System: Zivan NG1-DC
Wiring: Prestoflex #2/0 insulated welding cable
Switches & Contactors: Albright SW-200B Contactor
Fuses/Breakers: Airpax JLE-1-1-53-3-B4-250 main breaker; FERRAZ-SHAWMUT A30QS500-4 500-amp fuses; Painless 12-circuit ATO Fuse Center
Instrumentation & Gauges: Westach A2C5-28 50-150VDC Voltmeter; Westach A2C6-30 0-500 Amp Ammeter; Westach 3CTH7-5MAX analog tachometer
Drivetrain Modifications: Late model MG Midget transmission mounted lower than stock in chassis; late model MG clutch system, front disc brakes, and dual-acting rear drum brakes
Suspension Modifications: Aftermarket front and rear springs; 1/2" front anti-roll bar; reinforced main front box members; tubular steel rear subframe; late model MG steering column
Body Modifications: replaced front bonnet (hood) with 1959-61 Mk1 Sprite "bugeye" unit; modified transmission tunnel; various mods for mounting of battery boxes
Interior: Custom white ash dashboard; heated seat pads; 12-volt ceramic heaters
Charging Time: 8–10 hours
Range: 40+ miles (with nine batteries)
Top Speed: 70+ mph
Acknowledgments: "I would like to thank my wife, Kathy, for her support of my dream. I'd also like to thank Ken Koch, of KTA Services, for his guidance when I started this project. In addition, I would like thank Jimmy and his team for the wonderful paint job and Frank and his crew for the great powder coating."

Mounting all ten Trojan T-1275 batteries in the diminutive Sprite required creative planning early in the conversion process. In the end, three separate battery boxes were required: one under the front bonnet holding four of the batteries, one behind the seats in the cockpit for two batteries, and the one in the rear boot (trunk), shown here, which holds the remaining four batteries. Courtesy Mark Hayes.

A custom subframe was constructed behind the seats in the cockpit of the car and extended back into the rear trunk area. In addition to providing mounting points for six of the large Trojan 12-volt batteries, the frame also provided an attachment point for a roll bar, the rear bumpers of the car, and for a future upgrade to coil-over shocks. The subframe also acts as a conduit for much of the rear wiring in the car. The owner recessed the subframe as low in the bodywork as possible, to both help weight distribution and also to ensure that all the batteries would physically fit. Courtesy Mark Hayes.

The Advanced DC 8" motor was mated to a later-model Midget 4-speed transmission with an adapter from KTA/Electro Automotive. Unfortunately the flywheel had a 6-bolt mounting pattern, while the adapter hub had a 4-bolt pattern. In addition, the transmission pilot shaft interfered with the end of the motor shaft, even without the flywheel and hub installed. Electro Automotive corrected the mistake by sending new parts, but the owner still had to machine a new hub from scratch on a lathe. Note that the top 3.5" of the transmission adapter plate and transmission housing were cut off to allow batteries to be mounted above the transmission. Courtesy Mark Hayes.

A strong frame made from angle iron steel was constructed to house the front four batteries. Note the electric motor tucked neatly underneath. Courtesy Mark Hayes.

The sides and bottoms of all three battery boxes were lined with 0.25" thick black polypropylene with the seams welded together. The top covers were made from 0.5" thick polypropylene. The covers were secured in place with bars made of steel channel iron that were bolted to the battery box frames, holding the batteries securely in place in case of a rollover. Each box has AC fans that vent any out-gassing that may occur during recharging. Courtesy Mark Hayes.

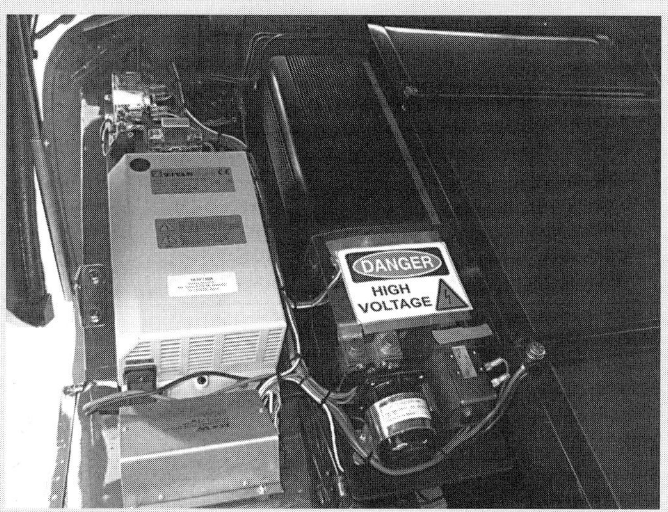

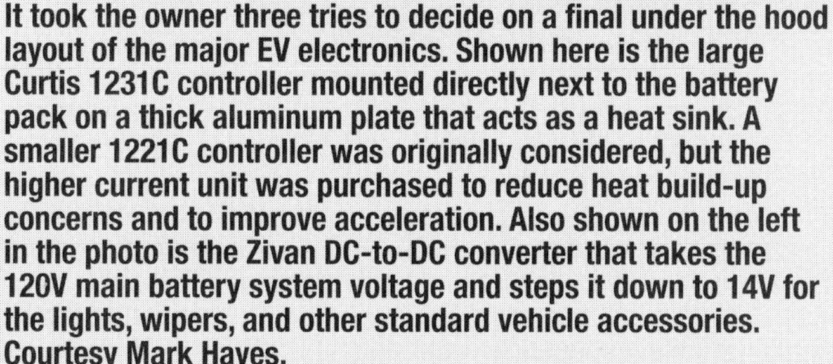

It took the owner three tries to decide on a final under the hood layout of the major EV electronics. Shown here is the large Curtis 1231C controller mounted directly next to the battery pack on a thick aluminum plate that acts as a heat sink. A smaller 1221C controller was originally considered, but the higher current unit was purchased to reduce heat build-up concerns and to improve acceleration. Also shown on the left in the photo is the Zivan DC-to-DC converter that takes the 120V main battery system voltage and steps it down to 14V for the lights, wipers, and other standard vehicle accessories. Courtesy Mark Hayes.

During the restoration process, the owner upgraded many of the original drivetrain and suspension components with parts from a later model MG Midget. For example, the tiny original front drum brakes were replaced in favor of much more reliable and powerful Midget disc brakes. This type of consideration is very important when converting any vehicle to electric power, as the batteries generally add considerable weight to most conversions. Having adequate brakes that are up to the job of stopping a significantly heavier car is a critical safety consideration. Courtesy Mark Hayes.

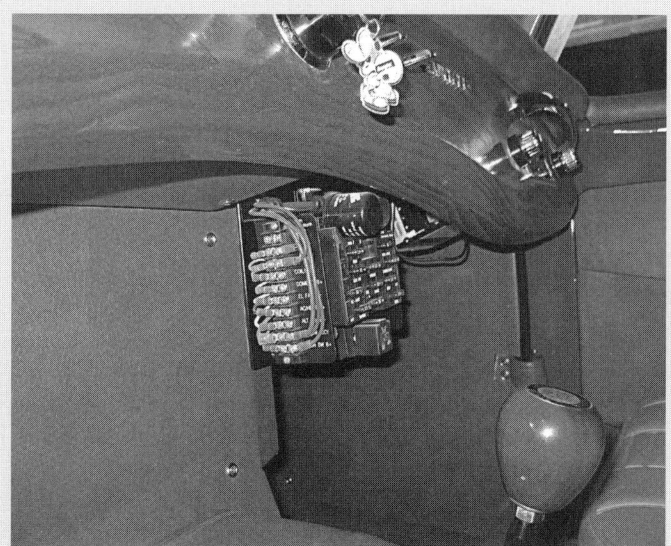

The owner used a Painless Performance Products 12-circuit ATO fuse center as the basis of the vehicle's electrical system. The fuse center is mounted to a slide that allows it to be pulled down for access, and then pushed back up under the dash where it is well protected and out of sight during normal vehicle operations. The owner wired and debugged each electrical circuit one by one, so that troubleshooting was easy to perform. Also, much of the entire EV conversion was debugged during early "prototype" phases and tests. This made the final installation and debugging process on the car very quick and easy, since most of the components and systems had been fully tested previously. Courtesy Mark Hayes.

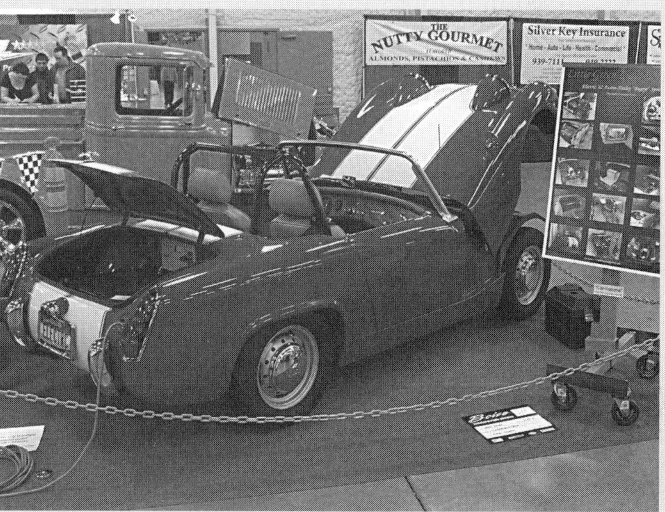

The completed car at a roadster show. Says the owner, "Every time I drive the car at least one person will honk and wave, giving me the thumbs-up sign and commenting on the car. Then, when they learn it's powered by electricity, they're usually amazed." The car has won several awards, including a first place and best-engineered awards at a 2009 roadster Show. The car also has been featured on all of the local news networks and in some of the local papers. Going green in a car like a Sprite is sure to get noticed. Courtesy Mark Hayes.

Project 4:

Porsche 914

With its manual steering, manual brakes, manual transmission, and air-cooled engine, the Porsche 914 is a nearly ideal candidate for conversion to electric power. The car is simple, lightweight, and has unique styling; what more could you ask for in a conversion donor vehicle? Potential cars should be fully inspected for rust and structural soundness prior to being purchased for conversion; these cars are nearly 40 years old, and some have been taken care of better than others. Note the 120V power cord plugged into the car for its nightly recharge. Courtesy David Oberlander.

In the late 1960s, Porsche and Volkswagen teamed together to create a lightweight, entry-level sports car that combined good performance, great handling, and unique styling. The original goal was to sell two variants of the car: one with a powerful six-cylinder engine and the other with a more economical four-cylinder power plant that shared its basic design with the then ubiquitous VW Beetle. While the six-cylinder model never became a hit with the general public, the four-cylinder car was an overnight success. Advertised as an "entry level Porsche," the little two-seater garnered praise from racers and consumers alike, and *Motor Trend* named the 914 as its Import Car of the Year in 1970. The Porsche 914 was one of those perfect examples of the right car at the right time for the right price.

Fast-forward 40 years. The Porsche 914 still enjoys cult-like status among knowledgeable automobile enthusiasts. With its inherent low center of gravity, guided missile-like handling, and breezy top-down motoring, the 914 is just as desirable today as it was back in the seventies. As a result, prices for the diminutive vehicle have risen in recent years, with some six-cylinder models selling for mid five figures and above. The more popular four-cylinder cars are also rising in demand, but deals can still be had if the buyer is willing to shop carefully and then pounce when the right car shows up for sale—which is exactly what the owner of the car shown here did.

"I was forwarded a link to Electro Automotive, a company that specialized in Porsche 914 conversions," says owner David Oberlander of Richardson, Texas. "Then, after a couple of weeks of searching, I found a running car at a great price just a few hours after the ad was posted. I purchased the vehicle over the phone, sight unseen."

Unfortunately for David, the car had a few problems, some more serious than others. "After I purchased the car, I registered and insured it, troubleshot the electrical system, and then drove it directly to the paint and body shop. During the short drive, the driver's side seat literally dropped through the rusted floor pan and I nearly skidded across the highway going 60mph!"

Needless to say, new floor pan steel was ordered and installed. During this time, David also placed the order for the major kit components for the EV conversion. In an effort to save money, he did not order a complete kit from Electro Automotive; instead he modified it by using 8-volt batteries in lieu of 6-volt, which reduced the weight and range of the vehicle. The kit components included motor and mounts, transmission adapter, controller, charger, two battery boxes, instrumentation, contactors, and detailed instructions. David works at Dodson Services, an electrical contractor in Texas, so he was able to order the wiring, batteries, relays, wire lugs, connectors, and other items directly from their electrical vendors, which saved him a few

The original air-cooled internal combustion engine (ICE) displaced 1.7-liters and generated around 80hp at the crankshaft. It was also loud, greasy, leaked oil, and produced more than its fair share of air pollution. Converting the car to electric power took approximately five months. Courtesy David Oberlander.

The electric motor chosen to replace the ICE was an Advance Motor Company FB1-4001A series-wound DC unit that can produce 85hp at 120V. Note the Curtis 1231C-8601 500 amp 144v on road motor controller mounted to the passenger-side firewall. Note, too, the Electro Automotive aluminum adapter plate used to attach the electric motor to the 914 stock transmission. A few months after installation, the electric motor failed. A factory inspection found an armature fault, and Advance Motor Company sent a replacement motor within a couple weeks, free of charge. Courtesy David Oberlander.

dollars as well. A couple of weeks later, the car came out of the body shop.

"I was a little nervous," says David, "seeing this pristine classic coupe, fresh from the shop, and knowing that I was about to butcher it. I had already invested a lot of money in the car, and the first thing I was going to do was permanently modify it."

David's boss, Drew Dodson, allowed the Porsche to be converted in the company's shop during the

SPECIFICATIONS

Vehicle: 1972 Porsche 914-4 targa-top convertible
Original Engine: 1.7-liter 4-cylinder horizontally opposed gasoline internal combustion
Motor: Advance Motor Company FB1-4001A 25.2 hp continuous 85 hp peak at 120V series-wound DC motor
Motor Adapter: Electro Automotive clutched adapter plate and flywheel hub assembly
Controller: Curtis 1231C-8601 500 amp 144V
Batteries: 15 8V 183ah lead-acid U.S. Battery US 8V-GCHC-XC
System Voltage: 120V
Charger: Russco SC 18-120 SO 15A 1800W 120V charger with digital shut-off; a two- to six-amp automatic charger is used for 12V auxiliary battery.
DC-to-DC System: Kelly HWZ Series 120V-12V 25A 300W converter from Cloud Electric
Wiring: Albright SW200 switches the 120V power to controller with a safety relay that engages the contactor when a micro switch on the 5K ohm throttle pot is engaged and when key switched On; 250A Airpax DC breaker located near the clutch to act as an emergency power off device and to serve as over-current protection for the 120V system; key switch circuit (KSI) engages a fan relay for battery box ventilation; two relays are used to activate the line and load side of the DC converter when the KSI is engaged; Interlock relay prevents any operation of the EV during charging.
Instrumentation & Gauges: Electro Automotive 0-500A ammeter; 105-135V voltmeter (state of charge); stock speedometer; custom-dash warning lights (KSI ON; motor overheat; charging, reverse gear); digital Kill-A-Watt meter to monitor charging and operation current draw.
Safety: Switched reverse beeper used in high-pedestrian areas (e.g., parking lots)
Drivetrain Modifications: Replaced clutch assembly, flywheel, throwout bearing, front seal, and synthetic transmission oil
Suspension Modifications: welded-in chassis stiffening kit; 180 lb/in rear coil springs; upgraded front and rear Bilstein gas shocks; low rolling resistance Yokohama YK520 195/65/R15 91H tires
Body Modifications: Marine charging port on passenger front quarter panel
Interior: Road Pro 300 watt 12V ceramic heater and fan
Charging Time: 16 hours for full charge (40-mile trip)
Range: 40 miles
Top Speed: 80+ mph
Acknowledgments: "I would like to thank Drew Dodson, president of Dodson Services Company. I could not have completed this project without his assistance. I'd also like to thank Bill Lentfer of Electro Automotive. He is the man who made available to me the incredible EV Porsche 914 conversion kit components with the highest quality parts and instructions."

Six of the car's 8-volt 183 Amp-hour U.S. Battery 8V-GCHC-XC batteries are mounted in a special battery box in the front of the vehicle. A cover plate (not shown) is used to close the box during normal operation. Total vehicle system voltage is 120 volts. Note the 120V charger centered above the battery box. The owner says that the 1800W charger is sufficient, but to reduce the charging time, a more powerful charger such as a 3000W model could be installed. Note the smaller 12V charger to the right that is used to recharge the auxiliary battery. Courtesy David Oberlander.

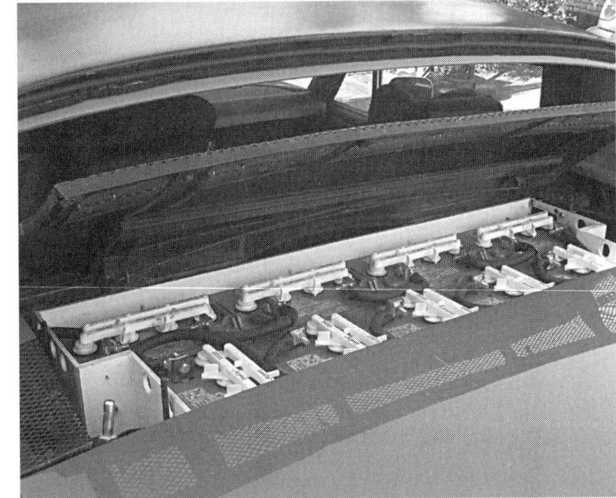

The other nine 8-Volt batteries are located in a special box at the rear of the vehicle, where the original ICE was located. Courtesy David Oberlander.

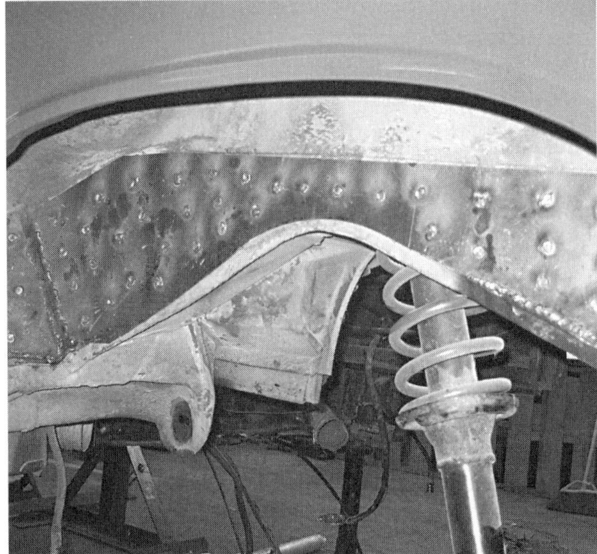

To help support the added weight of the batteries, a plate-steel chassis stiffening kit was welded to the frame. This type of kit is available from a number of aftermarket suppliers, but requires installation by a qualified welder. Courtesy David Oberlander.

summer of 2008. The car was lifted up onto jack stands, and then completely stripped of the stock gasoline engine, along with all other remnants of the ICE, including gas tank, fuel pump, heater system, and exhaust. A rear chassis stiffening kit was welded in place, and a large steel plate was mounted under the rear passenger firewall to be used to mount the controller on and act as a heat sink. Following this, the undercarriage and EV compartment were rust-proofed and painted. Then the more interesting part of the conversion began.

"By August of that summer, EV components were arriving in the shop. This really was the fun work for me. Mounting the controller, installing the contactor and throttle pot, coupling the motor to the transmission and getting it installed. Everything went together without a hitch." Well, almost everything. A challenge was discovered during the installation of the rear battery box steel support frame: it didn't simply set into place. "The battery boxes were well crafted units with precision mounting clearances, and the rear box support didn't easily slide into the engine compartment at first. I think it was because my car wasn't perfectly straight. Fortunately, I discovered the magic of using a pneumatic hammer to massage the chassis and squeeze the battery box into place."

Once the boxes were in, David focused on the interior of the 914, including the installation of a new steering wheel, replacement dash, new carpet, EV pillar gauges, seat covers, mirrors, window seals, latches, and so on. The final stage of the conversion involved installing the batteries and lowering the vehicle to test drive. Unfortunately, the weight of the car was too much for the worn-out stock springs, and the car simply sat too low for safe use. A set of heavy duty 180 lb/in rear springs, new gas shocks, and a readjustment of the front torsion bar raised the car back to factory height.

The latter part of October was spent on troubleshooting the wiring circuits. "I also spent this time to add new brake pads, and work on more loose ends. Finally, it was Halloween when I sat in the driver's seat, turned the key for the first time, pressed the accelerator, and heard the high-pitched sound of the Curtis Controller...and took off!"

The car was then tested by a Texas state vehicle safety inspection station and passed. David has subsequently tuned and tweaked the car, replacing tires, installing an electric heater, and so on. He now regularly drives the car to and from work with very few problems. Says David, "I enjoy driving the EV and continue to find ways to improve it. Working on the car is nearly as much fun as driving it."

Larger capacity (i.e., stiffer) springs were required at the rear to help carry the load of the batteries and restore the car to its normal factory ride height. These units are 180 lb/in coil springs. Stiffer springs were not required at the front of the vehicle, but the factory torsion springs did require re-adjustment to level the car. Courtesy David Oberlander.

The 5000 ohm throttle pot is mounted to the driver's side firewall. Note the dual (redundant) return springs on the unit; if one spring were to break, the other spring can still return the throttle to the safe off-position. The cable leading from the left of the swing arm is connected to the OEM accelerator pedal inside the cabin. Courtesy David Oberlander.

The 12V auxiliary battery is located at the front of the vehicle. Note the 120V-to-12V converter located on the right side of the image, next to the main 120V charger unit. Courtesy David Oberlander.

Cabin heat is provided by way of a small Road Pro 300-watt 12-Volt ceramic heater unit. Courtesy David Oberlander.

Two A-pillar mounted gauges are used to monitor battery system state of charge (upper gauge) and main system current flow (lower gauge). The speedometer and odometer are stock Porsche units (not shown). The owner has modified the stock alternator light to indicate when/if motor temperature is exceeded. Similarly, the fuel light switch now indicates disabled operation of the EV when the batteries are being charged. Courtesy David Oberlander.

Project 5:

Ford Ranger

A very clean and well-executed Ford Ranger EV conversion at home on the beach in Florida. Except for some small custom "Electric" decals on the exterior, most drivers on the road would never suspect that the step side pickup truck beside them is electric powered. The owner says, "The best part about converting an internal combustion engine car to electric is that you don't need to be an expert. I am not a mechanic by any means. What I learned about EVs I learned by doing it. Anyone can do this if they take their time to understand and learn." Courtesy Mike Sileo.

When Edward Reim of Tavernier, Florida, decided to build an electric vehicle, he knew he didn't want to compromise during the process. He wanted to build his own vehicle but not one that lacked safety features or creature comforts. He wanted economical electric power, but he also needed a car that could travel a relatively long distance and keep up with traffic on the highway. In short, Ed wanted his electric cake and he wanted to eat it, too. "One of the main reasons I did a conversion was to show people that you do not have to compromise safety to drive an electric car, and that any car can be converted with all the bells and whistles that it had to begin with. I wanted long range on a single charge to make it to work, and I wanted a relatively high top speed. My commute is 90 miles round-trip, and I didn't want to create excess carbon emissions during the drive, but I live in the Florida Keys, so I also wanted air-conditioning and all the standard power accessories."

Ed started the ambitious project by purchasing a 1996 Ford Ranger XLT step side pickup truck from an advertisement he saw in the local Auto Trader. Finding the car was the easy part; deciding on the specific electric power configuration was a little harder. "I spent nine months researching EVs and components on the Internet before deciding that an AC kit from Electro Automotive would best fit my needs. The AC system has built-in regenerative braking and is smoother on the transmission. It also uses less current than an equivalent DC system and should allow my batteries to last longer. My employer is considering adding power outlets in the parking lot for recharging, but I wanted a system that could make the round-trip with or without that option."

After the order was placed for the conversion kit from Electro Automotive, Ed began preparing a place at his house to tear down and then build up the truck. Florida may be officially known as the "Sunshine State," but it rains a lot there, too, especially in the Keys. Without a traditional garage to work in, Ed decided a temporary shelter would best suit his needs, so he ordered a 24-ft x 12-ft x 10-ft high tent from Shelters of America. Once the tent was assembled and interior lights installed, the EV conversion parts began showing up.

Ed started the conversion process by removing the truck's internal combustion engine (ICE) and all of its supporting equipment and subsystems. "I tagged everything that was removed with a handheld labeling machine. Then, with the help of my neighbor Frank, the ICE was pulled, which took about five hours and half a gallon of orange hand cleaner." Following this, Ed removed the last vestiges of the gasoline engine, including the fuel tank and exhaust system, and then went to work cleaning, rust-proofing, coating, and painting the frame and chassis of the truck.

The installation of the AC motor, controller, and other equipment was relatively easy and mostly went according to plan. There were, however, a few minor hiccups along the way. "I spent quite a bit of time hinging the bed of the truck with pivots and gas springs, with the idea to mount the batteries underneath the bed. I wanted to show that the vehicle could be converted without losing any of the functionality that the original vehicle had. Unfortunately, after building the hinges Frank and I realized that we would not be able to mount all 26 batteries underneath without stacking them. The driveshaft in a Ford Ranger is not centered symmetrically, which angles the driveshaft and prevented me from mounting the batteries the same on both sides. Other trucks, like the Chevy S10 have a centered driveshaft, but not the Ranger." He also began to have second thoughts about the stock vehicle rear axle carrying the entire load of 26 heavy batteries.

After some head scratching, Ed decided to split the batteries up, with five mounted under the engine bay hood, and the rest in the bed of the truck to help even out the load of the rear axle. "Even though I couldn't mount the batteries underneath, I'm keeping the hinged system. In the future, when smaller batteries or ultracapacitors are available to the general public, I intend to relocate them underneath. In the meantime, the current batteries work fine in the bed and under the hood."

The remainder of the conversion went smoothly, including installation of the controller, cooling fans, charger, and all the wiring and connectors. Ed is a self-admitted "neat freak," which explains the orderly engine bay layout and tidy electrical wiring. He even went so far as to build his own dashboard from wood, complete with truck-shaped cutouts for his turn signal indicator lights. An electric-powered air conditioner was also installed and plumbed into the vehicle's cooling ductwork. Then it was time for the moment of truth.

"The battery pack was wired and tested, with 330 volts measured at the input to the controller," Ed says. "We turned on the ignition switch and all the lights powered up. I then tried to move her under power for the first time." The car was placed in first gear and the accelerator pedal was depressed slightly. "Nothing happened, so I pressed a little harder...Still nothing, so I floored it and the car jumped backwards!"

It turned out that the car moved forward in reverse, and backward in first gear. Ed emailed Electro Automotive, who told him how to change the Shaft Direction parameter in the controller software. After making that one simple change, the car moved in the right direction. Ed then set out testing the vehicle and debugging some minor gremlins, including a non-functioning speedometer and a balky charger. Then it was a series of longer and longer test drives. Ed now routinely drives the truck to work with nary a problem. Asked if he likes his electric Ford Ranger commuter truck, Ed responds, "Absolutely. The range is fine and it easily keeps up with traffic. I pass gas stations without even bothering to check the price of unleaded. I know now what the EV Grin is all about!"

SPECIFICATIONS

Vehicle: 1996 Ford Ranger XLT pickup truck
Original Engine: 2.7-liter 4-cylinder gasoline internal combustion
Motor: Azure Dynamics/Solectria AC55 3-Phase AC Motor
Motor Adapter: Custom 4-piece adapter made by Electro Automotive as part of the kit
Controller: Azure Dynamics/Solectria DMOC445
Batteries: 26 12V lead-acid GEL MK 88AH 27T
System Voltage: 312V
Charger: Zivan NG5 220V 30A
DC-to-DC System: Azure Dynamics/Solectria DCDC750
Wiring: 2awg for long runs, 4awg for short runs between batteries
Switches: Regen on and off switch in cab, selector switch to choose between Power / Normal / Economy with respect to regenerative braking
Fuses/Breakers: Two fusible links. One in the front of the truck with 5 batteries and one in the back with 21 batteries; will blow if there is a dead short on the pack.
Instrumentation & Gauges: Custom Dash. Electro Automotive 300V voltmeter; +/- 200A ammeter; 16V voltmeter for accessories.
Safety: Relay that tells the controller when 240AC is present during a charge so the vehicle cannot move while the cord is plugged in.
Drivetrain Modifications: None
Suspension Modifications: Super Spring helper springs on rear to help carry load of batteries
Body Modifications: Hinged bed
Interior: No heater; 12V solid state air conditioner
Charging Time: 6-8 hours
Range: 75 miles
Top Speed: 95 mph
Acknowledgments: "I'd like to first thank my wife, Lisa, and the rest of my family for putting up with my time spent on this project and missing some key events in the process. I also want to thank Frank Murrell, for without him I would not have been able to complete this car. He was the fabricator and metal designer. We would talk about an idea in the morning and by the end of the day a painted piece was installed and working. I also need to thank my neighbor Steve Lavoie, my friend and photographer Mike Sileo, and all the guys at Turkey Point who put up with my stories and photos on a daily basis so I could work out problems and solutions."

The Azure Dynamics/Solectria AC55 3-Phase AC Motor was chosen in part for its built-in regenerative braking abilities. An adapter to mate the motor to the Ford transmission was provided as part of the Electro Automotive kit. Courtesy Ed Reim.

A view of the Ranger's "engine" compartment. Not visible is the Dynamics/Solectria AC55 3-Phase AC Motor, which is located underneath the large Azure controller in the center. Five of the vehicle's traction batteries are mounted at the front of the engine bay (three visible; two underneath). Note the single Bosch 12V battery (arrow) that is used to power the standard electric systems, such as lights, horn, and windshield wipers. Courtesy Mike Sileo.

Twenty-one of the vehicle's 26 12V batteries are carried in the bed of the truck. Note the steel frame used to hold down the batteries and keep them in place. The other five batteries are mounted under the hood at the front of the vehicle. Courtesy Mike Sileo.

The owner built his own custom dashboard from wood. From left to right, the gauges are: (a) a 200-amp ammeter for measure current flow to and from the battery bank; (b) a 300V voltmeter for measuring main battery system voltage; and (c) a 16V voltmeter for measuring auxiliary system voltage. Also shown is the aftermarket speedometer on the far right. Note, too, the custom truck-shaped turn signal indicator lights. Courtesy Mike Sileo.

The charging port is located in place of the original gas filler cap. "I am very pleased that I can drive a regular sized vehicle at the same speed or faster than a gas-powered vehicle and pass the gas stations without looking at them," says the owner. "This truck has freed me from worrying about daily oil price fluctuations." Courtesy Ed Reim.

The proud owner cruising on the road in sunny Florida. "The AC motor is very smooth and powerful, but it did take a while to learn how to drive the truck," Ed says. "The system has regenerative capability, which essentially turns the motor into a generator when slowing down, directing electrical current back into the batteries. The system works great, but I have learned to turn this feature off at high speed, as it puts incredible drag on the vehicle when slowing. I once made the mistake of taking my foot off the pedal at 45 mph to stretch, and it was like slamming on the brakes." Courtesy Mike Sileo.

Project 6:

Saturn

In 1996, General Motors leased its short-lived EV1 electric car through Saturn dealerships. Today, the only Saturn-supplied electric vehicle you can own is one that has been converted by an enthusiast from ICE power. The owner of the 1999 SL2 four-door shown here wanted an EV to counter rising gas prices. He chose a Saturn to convert because of its low coefficient of drag, its plastic body, and its low weight. His goal when beginning the conversion was a car that could travel 35 miles round trip to work on a single charge, reach at least 60 mph, and provide "maximum fun when driving past gas stations." The conversion has met all of the original goals, and has even won an award at a local car show in the Special Interest class. Courtesy Wayne Krauth.

In 1985, General Motors created the spin-off company Saturn to compete directly with the best-selling small imports made by Honda and Toyota. Saturn cars were built in the U.S., sold through an independent retail network separate from traditional GM dealerships, and were offered to the public at simple "no hassles, no haggles" prices. The results of this novel approach were considerable; sales of the early S and, later, the L-series vehicles were very strong, with buyers tending to come back for repeat purchases. Saturn vehicles may not have been particularly flashy to look at, but they had well-earned reputations for impressive reliability and sturdy construction. Also, their lightweight bodies were made largely from plastic composites, which offered excellent rust protection and allowed for efficient, lightweight and aerodynamic bodywork. In other words, Saturn cars have many of the intrinsic characteristics that EV enthusiasts look for when considering donor vehicles. For these reasons, Wayne Krauth of Standish, Maine chose a 1999 Saturn SL2 sedan for his first conversion project.

"In 2008, gasoline was at four dollars per gallon," says Wayne. "I was unhappy with the rising cost of gas and record oil company profits, especially when coupled with the tax breaks they were given. Therefore, I decided to go electric. For my conversion, I wanted either a popular sedan or light truck that was American made. I also wanted something with good aerodynamics and lightweight. Living in Maine also meant trying to find a car with little to no rust. The plastic-bodied Saturn fit all of these criteria. Saturns were available locally for sale in the newspaper, and I could buy parts at just about any auto parts store."

Wayne found a four-door SL2 nearby for sale with a blown engine. The rest of the vehicle was otherwise in good shape, so a deal was struck. The Saturn was then towed home to Wayne's house and the conversion process started.

"I have a background in electronics and a strong mechanical aptitude. I also enjoy a good challenge. I am not, however, an experienced auto mechanic, nor had I ever tackled an EV conversion before. Once I had narrowed my choice down to the Saturn, I looked at previous conversions I found on the EValbum.com website to see what others had done. This helped me determine key parts, the overall battery configuration, calculate things like potential range, and ensure the viability of my choices."

As a result of this research, Wayne decided that a basic lead-acid battery and DC-motor system would be straightforward to tackle and would meet his range goal of a

The motor selected for the conversion was an Advanced Motors and Drives FB1-4001A 9" diameter, series-wound DC motor. The owner considered an AC system because of its higher efficiency, but a DC motor was deemed simpler and more affordable. The four-piece KTA adapter plate and coupling is shown here attached to the front of the motor prior to installation in the car. Courtesy Wayne Krauth.

The owner shown lowering the electric motor down into the engine bay by way of an overhead hoist. Says the owner, "Bolting the motor into the car is a huge project milestone, especially when building your first EV." Courtesy Wayne Krauth.

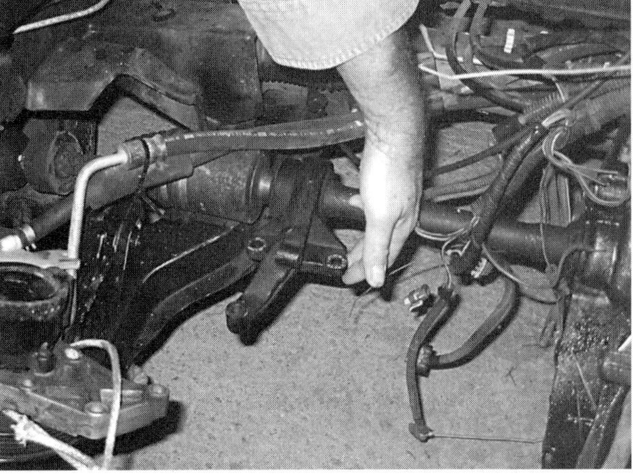

Many front-wheel drive vehicles utilize an intermediary driveshaft that connects the passenger side half-shaft to the transmission (transaxle). The bearing bracket that is being supported by hand in this photograph holds the outer end of the Saturn's intermediary shaft. After the electric motor was installed, a steel adapter plate was added to the outboard motor mount that interfaced to this bearing bracket, thereby providing permanent support. The owner says that deciding on, and implementing, a suitable mounting scheme for the bearing bracket was the most difficult task encountered during the conversion process. Courtesy Wayne Krauth.

round-trip commute to and from work on a single charge. It would also be affordable, which was a key consideration. Says Wayne, "DC-based systems are more common and less costly. AC systems are more energy efficient, but also more expensive." For his conversion, Wayne selected an Advanced DC series-wound 9" motor with a Curtis controller.

A similar trade-off of weight versus range versus cost was applied to his choice of batteries. "Flooded lead-acids were the least expensive," says Wayne. "They also have proven reliability and could be purchased easily. The down side is they're heavy and require maintenance. I live in Maine, too, so I worried a bit about their reduced capacity in cold weather. Lithium batteries can handle the cold much better than lead-acids. They also would have been a much lighter choice, and would require much less maintenance. The range you can achieve with them is also considerably higher, because they can tolerate a deeper discharge than lead-acids with less effect on their lifespan. The downside, of course, is lithium batteries would have cost a lot more money and required an expensive battery management system."

In the end, Wayne decided to use fifteen Trojan 8V flooded lead-acid batteries, for a total system voltage of 120V. "The 8V batteries I selected had the highest capacity in that type of Trojan batteries. This allowed me to have a 120V pack voltage, keep the total weight of the vehicle well under its GVWR, and maximize its potential range."

Once he had selected the basic system components, it was time to perform the actual conversion. The internal combustion engine and all its supporting equipment was removed. Battery locations were then determined and support frames constructed from angle iron. The next step was to mount the motor to the transmission. Unfortunately, this wasn't quite as straightforward as simply making an adapter plate and bolting the two together. Like many front-wheel drive vehicles, the Saturn SL2 uses an intermediary half-shaft on the passenger side, which connects the drive halfshaft to the transmission. In the original ICE configuration, the intermediary shaft is supported by a pilot-bearing that is bolted to the internal combustion engine. With the ICE removed, Wayne had to recreate this bearing mount on the electric motor.

"Making the mount for the intermediary half-shaft, and then aligning it precisely was the most difficult problem I encountered during the conversion project. I handled it by first mounting the motor firmly at both ends and then positioning the half-shaft in place, tack-welding its mount in position, rechecking everything, and then removing the motor and shaft mounting assembly so I could fully weld it up. After this, the rest of the assembly and conversion work went smoothly. It was just a matter of finding a reasonable spot for the batteries and everything else that needed to fit under the hood, and then building mounts and brackets for them all."

After everything was installed, the front of the car was within 40 lb of the original weight, so no changes were required to the front suspension. The rear of the car, however, was a different story. To help carry the estimated 600 extra lb load of the lead-acid batteries at the rear, custom heavy duty springs and heavy duty struts were installed, which brought the car back to original ride height. Then it was on to final wiring and testing.

Wayne says that he performed significant testing on individual parts and circuits as they were installed in the car. Consequently, the system worked as expected the first time it was powered up. The only exception was a faulty speedometer, which was traced to a bad ground wire. Otherwise, the Saturn was surprisingly trouble-free.

"When I first decided to do the conversion," says Wayne, "I wasn't really sure I would be able to complete it on my own. In hindsight, it was not nearly as bad as I initially expected. Just take it slow, doing a bit at a time and thinking things through before doing them. Think about what comes next and how that step might work with or interfere with remaining construction tasks."

Wayne also says that now that the vehicle is running, it's been great fun to show it off to friends, relatives, co-workers, and strangers alike. "Everybody wants to see how it works, and they all have the same questions. How fast? How far? What's the charge time? What parts are used?"

They also ask what the car is like to drive. Wayne reports that the Saturn operates like a cross between an automatic and a manual. "I don't use first gear much. To start from a stop, I use second gear but do not use the clutch. I simply press the pedal and start to move." He does, however, use the clutch to shift up to third (at 25–30 mph) and fourth (at about 40–45 mph). When slowing down, he just applies the brakes and can come to a stop without using the clutch. "At first you expect the car to buck and stall as you slow, but electric motors work fine all the way down to zero rpm, unlike a gasoline engine. The Saturn is smooth as silk to drive."

SPECIFICATIONS

Vehicle: 1999 Saturn SL2 four-door sedan
Original Engine: 1.9-liter 4-cylinder gasoline internal combustion
Motor: Advanced Motors and Drives, FB1-4001A, 9" diameter, series-wound DC motor, single shaft
Motor Adapter: Four-piece adapter purchased through KTA services
Controller: Curtis 1231C-7701
Batteries: 15 Trojan T890 8V flooded lead-acid
System Voltage: 120V
Charger: Zivan NG3 115 VAC @ 15A
DC-to-DC System: Zivan NG1, input 120 VDC, output 14.1 VDC, 50A
Wiring: 2/0 gauge welding wire to carry main power; 250A main circuit breaker; 500A fuse
Instrumentation & Gauges: Dash: Curtis battery gauge; 150V voltmeter; 500A ammeter; battery pack thermometers. Under hood: digital voltmeter and vacuum gauge.
Safety: Inertial switch to kill main power in event of collision. Main Contactor set up to engage only if: 1) the ignition key switch is on and 2) the car is not plugged into AC charging power; 3) the inertial switch not triggered and 4) the throttle pedal is pressed.
Drivetrain Modifications: None
Suspension Modifications: Heavy duty springs and struts installed on rear to help carry load of batteries
Body Modifications: Aerodynamic belly pan underneath car
Interior: 120V ceramic in-cabin heater
Other: 35 watt custom battery heaters
Charging Time: 8-9 hours
Range: 47 miles at 80% battery depth of discharge
Top Speed: 75 mph
Acknowledgments: "I would like to thank KTA Services for supplying parts, excellent schematic diagrams, answering my questions and responsively replacing a failed part. I'd also like to thanks Joe Labrecque for his hands-on help and advice."

Aside from the pure pleasure of driving, the real icing on the cake for Wayne is the dramatically reduced fuel costs. "The price of electricity is far less than gasoline," says Wayne. "For my daily commute, and including weekend trips, the total cost is about four to five dollars per week for electricity. I tell people that the choice is easy: four bucks a week in electricity, or four bucks a gallon in gas. It's a no-brainer!"

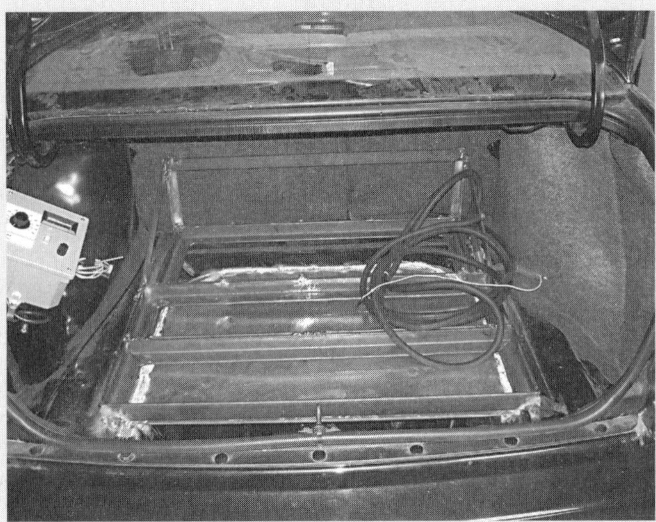

Steel brackets were constructed from angle iron and then installed in the rear trunk to hold the majority of the vehicle batteries. The total time required to convert the car from gasoline power to electric was 210 hours. Says the owner, "A lot of this time was spent learning skills like welding and metal working. I also spent a lot of time just designing the overall layout and details of the system. If I were doing it again, the next EV would probably take me half the number of hours." Courtesy Wayne Krauth.

The rear trunk shown filled with the Trojan 8-volt flooded lead-acid batteries. Total system voltage is 120-volts. The size and number of batteries were chosen primarily to achieve the desired minimum range of 40 miles. Note the small gray electronics box on the left side of the trunk, which is used to control the heaters that sit underneath the batteries. Courtesy Wayne Krauth.

A photograph showing the construction of the battery heaters. Each battery sits on one of these heaters, which consists of (from bottom up): a fiberglass sheet, a heating element made from resistance wire, and an aluminum plate to transfer heat to the batteries. The heaters are 35 Watts each, and are all wired in parallel to a temperature controller mounted in the trunk. Because the car is operated in Maine, heaters are required in winter; the useful range from a battery can drop significantly in cold weather. Courtesy Wayne Krauth.

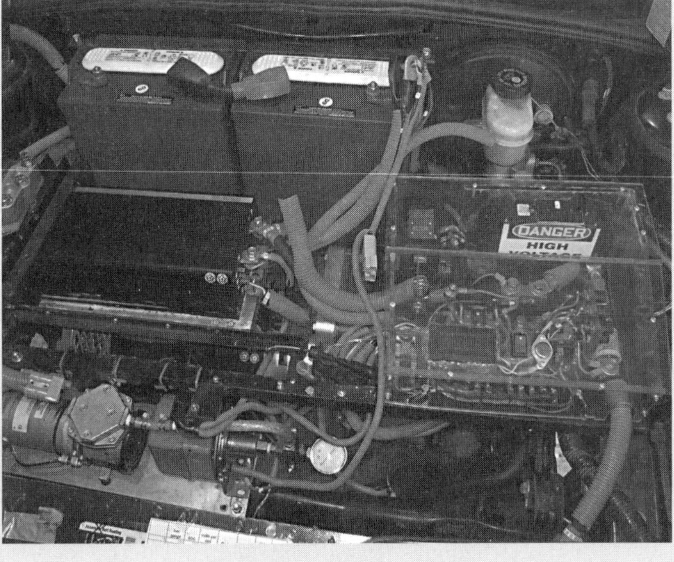

A view of the engine bay, showing two of the Trojan 8-volt batteries mounted at the front of the vehicle, along with the Curtis 1231C-7701 controller and some of the miscellaneous system wiring. In the lower left corner of the photograph is the GAST vacuum pump, gauge, and canister, which are used for operation of the power brakes. A vacuum switch serves to automatically turn the pump on and off, depending upon the amount of vacuum in the canister. Approximately 4-5 brake pedal cycles can occur before the pump kicks in to pull more vacuum in the system. Simple, but effective. Courtesy Wayne Krauth.

Another view of the completed engine compartment. The owner says, "I am pleased that everything works as it should. If I had to do it over again, however, there are a few things that I would change. The most important things are power steering and lithium batteries. I removed the power steering pump from the car, with the steering box hydraulic hose looped back onto itself. When the car is stopped, steering is heavy and requires two hands, but once it's rolling, steering is easy. For the batteries, lithium would have given me a much longer operational range. The trade-off, of course, is cost." Courtesy Wayne Krauth.

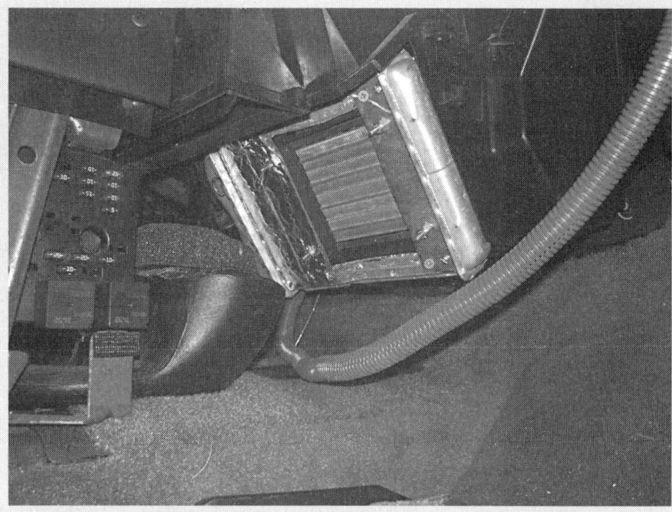

A simple 120-volt DC ceramic heater core was installed in a modified section of the original heater vent system underneath the dashboard. The unit is controlled by the normal fan switch, and it features self-limiting temperature control. Again, simple but effective. Courtesy Wayne Krauth.

Three EV system gauges were added to the dashboard: a Curtis battery gauge on the right, a 500-amp ammeter in the center, and a 150-volt voltmeter on the left. Under the hood, there are a digital voltmeter and vacuum gauge. The owner also applied adhesive-backed temperature-indicating "dots" to a number of underhood locations, such as on the motor and the controller. These stickers are designed to change color if a specific threshold temperature is reached, thereby giving the operator a means of telling whether certain components have overheated or not during use. Courtesy Wayne Krauth.

The Saturn plugged into a 115-volt AC 20-amp garage circuit for overnight charging. The charger is a Zivan NG3 that is custom configured for the vehicle's battery pack. Charging occurs in three distinct phases: initially with a steady high current applied, then with gradual reduction in current, and then finally with a low current "float" charge. Courtesy Wayne Krauth.

Project 7:
BMW Sedan

A 1995 BMW 325is prior to conversion to electric power. The goal was to create a reliable commuter vehicle that had good acceleration, a 65mph top speed, and at least 30 miles of range. The conversion was also a father-son project, and was intended to dispel the "glorified golf cart" image that most electric vehicles have with the general public. The owner chose the BMW primarily because it was small, well built, had good parts availability, and when finished would be a car he'd want to drive. Courtesy Rob Nicol.

Rob Nicol, of Corvallis, Oregon, is what you might call a serial EV owner. His first electric car was a Ford Cortina that he converted in high school using scrounged parts and an old aircraft generator for the motor. Later in college, Rob built an electric Honda Civic as part of a senior design course. The car was then used as a daily commuter vehicle that took Rob to and from work for the next five years. He then sold the Honda and later purchased a 1983 BMW 320i that had been converted to electric power by its previous owner in Seattle. Like the Honda, the BMW was driven daily by Rob for many years of trouble-free service.

Unfortunately, the car was involved in a serious motor vehicle accident in the summer of 2006, which left the car body and chassis destroyed. The good news was that there wasn't any spillage from the flooded lead-acid battery system. Even better, the electric drive system components were undamaged. Since Rob knew that another electric vehicle was in his future, he salvaged the EV parts from the wrecked car and went about searching for the next vehicle to convert.

"I knew I wanted another electric vehicle," says Rob. "Having the motor and controller salvaged from the wrecked BMW meant that I could convert another car fairly inexpensively. It also meant that some of the decisions and trade-offs were already made for me. For example, with the parts I had, I knew the system would need to be 144 volts. Because I wanted to use Optima AGM batteries, this meant twelve 12V batteries."

Rob narrowed his search of donor vehicles to two small, but different cars: a Mini Cooper or a 1993–98 BMW 3-series. "I wasn't strongly attached to either; whichever one I found first would be the car," says Rob. After six months of shopping online and watching the want ads, Rob located the perfect car, a 1995 BMW 325 that had a bad engine, but was otherwise in excellent shape. "My wife spotted the car on Craigslist. It was a couple hours drive away from our house, so my son and I borrowed a pickup truck and tow dolly from a friend, and we drove out to see the car."

Rob ended up purchasing the car and towing it home. His goals for the project were fairly standard for an EV conversion. He wanted good acceleration and excellent drivability. Range wasn't a major factor, as his commute is relatively short. Top speed, however, was important, because part of Rob's drive to work was on a 55 mph highway. There were also less tangible goals that were equally important. "I wanted to build the car with my youngest son, Greg, as a father-son project. I also wanted to build something that would be well-received by others once it was running. I wanted to show that an electric car could be more than a glorified golf cart."

After purchasing the car, the next step in the project was the gathering of all the components and parts needed to carry out the conversion. Much of these were salvaged from the previous car, but there were also a number of items that

One of the owner's previous electric vehicles was a 1983 BMW 320i that was, unfortunately, destroyed in an accident. The good news was that no one was hurt, and most of the electric components, including motor and controller, were undamaged, allowing them to be salvaged for transplant into the next BMW. Says the owner, "Since most of the electric drive system was removed and installed in the next car, this may be the only home built electric car in existence that has been previously crash tested!" Courtesy Rob Nicol.

SPECIFICATIONS

Vehicle: 1995 BMW 325is two-door sedan
Original Engine: 2.5-liter 6-cylinder gasoline internal combustion
Motor: Advanced DC FB-90011 9" DC motor
Motor Adapter: Custom, machined aluminum
Controller: DC Power Systems Raptor, 144V 600A
Batteries: Twelve 12V Optima absorbed glass mat, sealed lead-acid
System Voltage: 144V
Charger: Zivan NG5 220V (garage); Bonn 120V onboard charger
DC-to-DC System: BRUSA 14V, 250 Watt
Wiring: 2/0 primary cables; 400A fuse
Switches and Contactors: Albright SW-200B contactor
Instrumentation & Gauges: Standard BMW instrument panel
Drivetrain Modifications: Modified clutch
Suspension Modifications: None
Body Modifications: Projector headlights with Angel Eyes
Interior: Two 1500-watt ceramic electric heaters
Charging Time: 5 hours
Range: 30 miles
Top Speed: 75+ mph
Acknowledgments: "I would like to thank Otmar Ebenhoech, who was a great source of advice, encouragement, and loaned tools. Doug King provided many custom machined parts based on my poor sketches, plus consistent enthusiasm for the project. My son Greg did the heavy lifting, dirty jobs, and undercar work without complaint, and my wife Diane was a consistent and patient supporter of the project."

had to be upgraded or replaced. Motor adapter, wiring, and battery mounting hardware, for instance, all had to be sourced and/or fabricated. Says Rob, "I placed a large order with McMaster-Carr for items like high-quality battery terminal lugs, insulating covers and so forth. I also purchased a nice variable-pressure vacuum switch so I could dial in the boost on the brakes. It's in details like this that many home conversion projects fall down, in my opinion. Quality components and good planning are important to success."

After stripping the car of all its ICE-related equipment, Rob and Greg began figuring out how to "fake out" the BMW electronics. "Modern cars," says Rob, "especially the nicer ones, have a lot of sophisticated electronics and monitors that let the driver know if anything is wrong. For instance, the BMW has an alpha-numeric display for the diagnostic computer that tells the driver things like coolant level is low." Since the EV wouldn't need coolant, the reservoir and sensor were removed. Rob needed to determine what type of sensors were used (i.e., resistive or capacitive), measure what constituted a "normal" value, and then replace the sensor with the appropriate resistor or capacitor to make the computer think it was receiving the correct signal from a full reservoir.

One of the first things to do on a conversion project is remove what you don't need from the new car. This includes all of the equipment used to support the engine, such as gas tank, radiator, exhaust, plumbing, and electronics. Says the owner, "Some of the old parts could be sold, while others were scrapped. It helps to know ahead of time what is worth keeping because things can be removed faster if one doesn't mind destroying them." After everything was removed from the BMW, it was time to clean. "Once all the grease and dirt from the original ICE is removed, electric cars won't get your hands dirty working on them. We spent a lot of time and effort getting everything as clean as possible, because it is easier and more pleasant to work on a clean platform than a dirty one." Courtesy Rob Nicol.

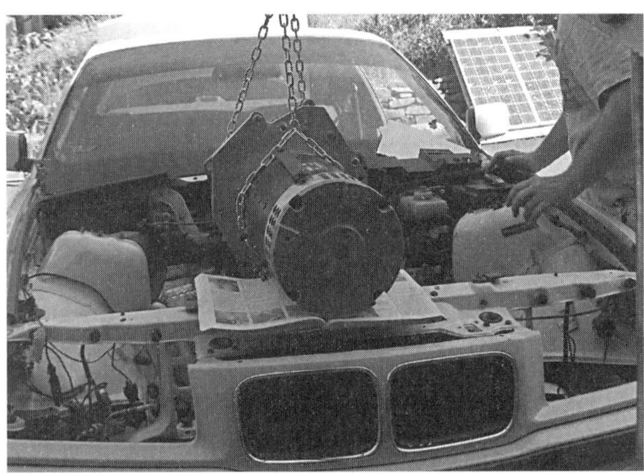

The Advanced DC FB-9001 9" motor and adapter being trial-fitted into the car. The owner was able to modify the old BMW motor-to-transmission adapter for use on the new BMW. Even though they were different generations of the 3-series BMW, the spline shafts on the bellhousing mounting holes were the same. The owner also used the clutch from the older car because it was lighter and the flywheel gear teeth had already been machined off. Courtesy Rob Nicol.

The next steps in the conversion involved layout and fabrication. In other words, where to put all of the batteries and components. "Some choices were easy," says Rob. "Half the batteries up front, half in the trunk between the wheel wells. Other decisions were already made for us. For instance, the motor would be bolted to the transmission, so the only real decision was how to adapt the two together and support the weight of the motor. Everything else took a lot of visualization and trial and error. Three or four different battery layouts up front were tried, for instance. We had to create strong brackets that avoided the steering column, and we also wanted to end up with short runs for high-current wiring between batteries." Like many other builders, Rob and his son utilized cardboard and wood mockups to test fit the equipment prior to building the real brackets out of aluminum or steel.

Once all of the brackets and adapters were built and all of the remaining equipment had been delivered, Rob and Greg began installation. This process went very fast, primarily because most everything had been test fitted and trial assembled. Final wiring was then carried out. The main power system was built using 2/0 welding cable that was connected with crimped copper lugs. Heat shrink tubing was used on all joints, and then red and black insulating covers installed. Components like the DC-to-DC converter, heater, and power steering pump were installed with 8- and 10-gauge wire. Smaller items, such as sensors and low-voltage relays, were connected with light gauge wiring. "To the extent possible, we tested everything as we connected it up. High current circuits were checked two or three times. We didn't want any surprises when we were finished."

The motor bolted to the transmission in the engine bay. "I chose to keep the original transmission because it was the easiest way to get gear reduction, plus it offered the flexibility in choosing different gears, as well as a mechanical rather than electrical reverse. In practice, I normally just leave the car in third gear for most driving. Occasionally, second gear is used for hills." Note the steering shaft on the right side of the photograph. The owner said that building battery brackets and mounts that were strong, yet didn't interfere with the shaft, was challenging. Courtesy Rob Nicol.

All of the planning, trial fitting, and testing during installation paid off the first time father and son turned the switch. Everything worked as it was supposed to. "I warned Greg that there are almost always errors and problems that you don't discover until you try the system the first time, but our incremental debugging during assembly worked flawlessly. Since there was nothing apparent that was faulty, we backed out of the driveway and took a short test drive. Nothing was found wrong, and I started driving the car daily from that point on." The only problem that surfaced occurred eight months later: a broken motor mount that was result of the owner answering the question: "How fast is it?" New, beefed-up motor mounts have since been fabricated and installed.

Rob reports that he is very pleased with the car, which can be recharged in five hours, driven thirty miles on a single charge, and has been taken up to 75 mph without any problems. Says Rob, "The choice of a BMW 3-series worked very well. The car is efficient, fun, and it looks good. It's heavier than it needs to be, but since where I live is flat, it's not a big deal. If I ever do another conversion, I will use a larger DC-DC converter to provide more margin when operating the lights, heater and wipers." Rob also says that he has plans for upgrading a few subsystems, including electric-powered air-conditioning, more advanced instrumentation (e.g., state of charge and energy usage/efficiency meters), and possibly lithium batteries, which would likely increase the vehicle range by a factor of four. For now, however, Rob is content just to drive and enjoy the BMW, which says a lot about both the car and its serial EV owner.

One of the more involved tasks was replacing the original vehicle heater with an electric equivalent. This job required a near total disassembly of the interior, and the fabrication of a heater core and controller. One benefit of electric heat is that there is no warm-up period; the heater begins producing warm air immediately after it is switched on. Courtesy Rob Nicol.

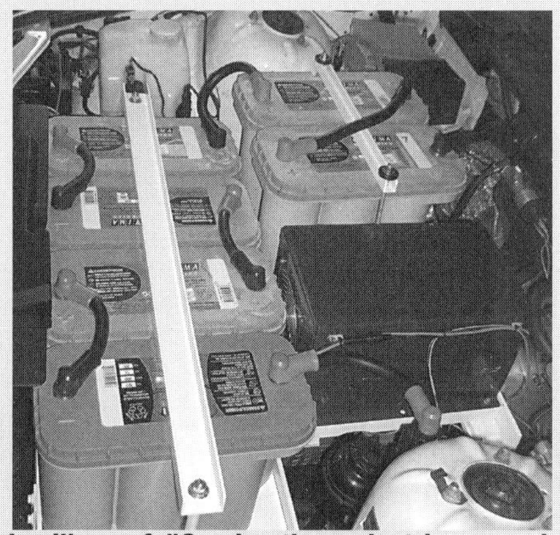

Twelve 12V Optima D31T yellow top batteries were used in the conversion, with six in the trunk, and six up front under the hood (shown here). The Optimas are spiral wound, valve regulated, absorbed glass mat, sealed lead-acid units that are maintenance-free and spill-proof. "Owning three electric cars prior to this one has provided a perspective on the capabilities and limitations of the technology," says the owner. "Batteries are definitely the limiting factor, but that doesn't mean a battery-powered car can't be practical for daily, local travel. I commute to work every day with my EV, and have far fewer mechanical problems than I would with an equivalent ICE-powered vehicle." Courtesy Rob Nicol.

Overnight charging is performed via a Zivan NG5 220V wall-mounted unit in the owner's garage. In addition, a Bonn 120V on-board charger located in the trunk is used for occasional opportunity charging. Range is a respectable thirty miles on a single charge. Courtesy Rob Nicol.

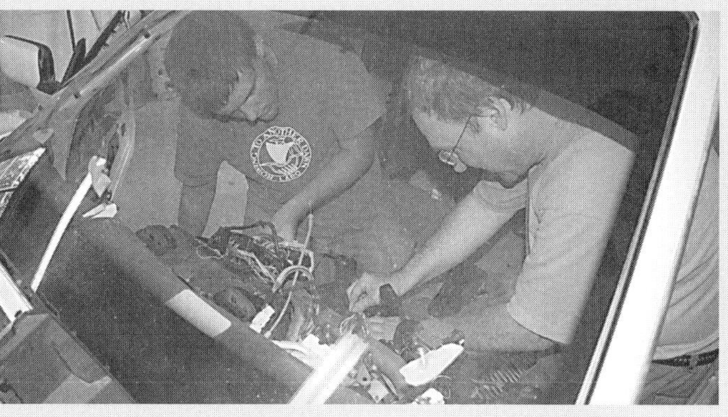

Working in their spare time, it took the father-son team approximately ten months to convert the 1995 BMW to electric power. Courtesy Rob Nicol.

The six Optima batteries located in the trunk at the rear of the car. These types of batteries are relatively lightweight and can be mounted in any orientation. Dividing the battery bank into two parts (one front, and one rear) ensured that no suspension modifications were required. Battery trays and hold-downs were fabricated from aluminum and painted white. They were then secured to the body with rivet nuts. Threaded rods were then dressed up with plastic tubing for a clean and professional look. Courtesy Rob Nicol.

The completed electric BMW is driven daily on the streets of Corvallis, Oregon. The owner reports brisk acceleration and a top speed in excess of 75 mph. Handling is also very good, thanks in part to the lightweight Optima batteries and the equal front-to-rear weight distribution. If it weren't for the "Electric" lettering on the side of the car (and the custom license plate) passerby's would be hard-pressed to tell the car is powered by electricity. Courtesy Rob Nicol.

Project 8:

VW Van

A 1978 Volkswagen Type 2 van converted to electric power. Air-cooled VWs like this are well-suited for conversion to electric power, as they are relatively lightweight, have a large load carrying ability, and are easy to work on. Adaptor plates for mounting a variety of electric motors to the transaxle are readily available, too. Courtesy Fred Weber.

Few things say "tree hugger" like an early Volkswagen Type 2 van. In contrast to mainstream American cars of the day, fitted with their high-power, high-tech engines, and sleek aerodynamics, the original VW vans were boxy and slow, with quirky air-cooled engines mounted at the rear, and cavernous interiors better suited for use as a work van or attending peace rallies than performance. More telling, VWs were frequently spotted wearing bright custom paint schemes adorned with flowers, peace symbols, and socially aware bumper stickers. VWs may have been practical to own when they first appeared on showroom floors, but the general public gravitated toward flashier automobiles, leaving the majority of sales of VWs "box on wheels" to freer spirits, broadminded activists, and independent thinkers of the time.

Even today, the typical owner of a Type 2 tends to be the unconventional type. Many of these folks are hands-on enthusiasts who care deeply about the environment and their impact on it. Enter Fred Weber, a computer chip designer who lives in northern California with his wife, Kate, and their two children. Fred and Kate are concerned about America's dependence on foreign oil and the related security costs and pollution created. They decided to zero out their daily driving carbon footprint with a rooftop solar electric system on their house and an electric car. First they looked at commercially available electric vehicles like the Zap Car but decided a home conversion would be more fun, more practical for daily use, and easier to maintain. Just as importantly, Fred was looking for a hands-on technical challenge to undertake. So, after researching the various options and products on the marketplace, they settled on a 7kW photovoltaic system on the roof of their house, and decided to convert a Type 2 van to electric power.

"I had a number of reasons why I wanted to build an electric vehicle," says Fred. "I wanted to lower my carbon footprint and make a statement encouraging others to look at electric cars, but I also wanted to just have the fun of building something innovative, and to learn firsthand how an electric car works." Kate also had good reasons for an EV, wanting an EV for daily use around town taking the kids to school and running errands, including carrying large groups of people. The couple also wanted to be able to bring home building supplies and such from the store. Maximum range was also important, but so was being able to go on highways when necessary. Says Fred, "A VW van was a great choice for me because it is large, has a high load capacity, and is easy to work on."

Fred originally wanted an early style 1967 split-window bus, but quickly discovered that the cost of a donor vehicle would be too high. In addition, the brakes and other mechanical subsystems on the earlier vans would have required significant upgrades to handle the heavy loads and forces generated by an electric conversion. Instead, Fred decided to look for a later model Type 2 van that had larger and more modern brakes. Then it was time to roll up his

The owner reports that one of the biggest decisions he faced early in the conversion process was between AC and DC power. In the end, he chose DC power and elected to use a Netgain WarP 9 series-wound DC motor due to its simplicity and low cost. Courtesy Fred Weber.

sleeves and learn everything he could about electric conversions.

"I found Mike Brown at www.electroauto.com and took his one-day seminar on electric car conversions. The class gave me valuable background information and confidence to do a conversion by myself. It also helped confirm my choice of using a VW van as a donor."

It took Fred about a month of looking at dozens of vans for sale before settling on the 1978 Type 2 shown on page 130. The vehicle had 69,000 miles on it, and, with the exception of some exterior surface rust and worn out tires, was generally in good shape. The next few months were then spent waiting for delivery of conversion parts, removing rust from the body, restoring the van's mechanical systems, and planning the EV installation.

"The key design decisions I faced in the beginning were the layout of the batteries, the design of the battery boxes, and the design and placement of the main control board. I went through quite a few iterations before finalizing the layout, but I'm very happy with the final design."

Fred divided the batteries into two groups. Fifteen of the twenty-four 6V batteries were located in a box directly behind the driver and front passenger's seats. This allowed Fred to keep the original rear seat in place with no modifications. The other nine batteries were located in a box that was sunk into the rear platform above where the original gas tank and engine were located. This created a flat interior platform for carrying cargo. It also helped fore-aft balance of the weight and provided reasonably safe

SPECIFICATIONS

Vehicle: 1978 VW Type 2 van
Original Engine: 2.0-liter air-cooled 4-cylinder gasoline internal combustion
Motor: Netgain WarP 9 series-wound DC
Motor Adapter: Custom Adaptor and taperlock hub by electroauto.com
Controller: Curtis/PMC 1231C
Batteries: 24 US 125XC 6V flooded lead-acid batteries
System Voltage: 144V
Charger: ElCon 110VAC/220VAC 1500W HF/PFC
DC-to-DC System: TDC 144V-12V
Wiring: 2/0 welding cable for main runs; 1"x1/16" copper strap for battery interconnects; Allbright #SW200 contactor; Airpax 500-Amp circuit breaker
Instrumentation & Gauges: Custom Dash. 500A ammeter with shunt; 144V state of charge (SOC) meter; 12V voltmeter for auxiliary system
Drivetrain Modifications: None
Suspension Modifications: KYB shocks; electric vacuum pump and reservoir for power brakes; Yokahama Y356 185R14/D1 tires at 60 psi
Body Modifications: None
Interior: No heater or air conditioner; stereo with iPod and Bluetooth
Charging Time: 12 hours
Range: 55 miles
Top Speed: 60 mph
Acknowledgments: "I would like to thank my wife, Kate, and daughter, Julia, who helped throughout the project. I'd also like to thank Mike Brown, who supplied the conversion kit and put the motor, adaptor, hub, flywheel and clutch together. And I'd like to thank my brother-in-law, Jeff Zurschmeide, who helped in the final two-day marathon to get the conversion on the road."

The main control electronics and switch gear is mounted to an aluminum plate that is suspended over the motor in the engine compartment at the rear of the vehicle. The owner says, "The goals were to keep everything as simple as possible. I also wanted to have enough range to use it as our primary vehicle and enough top speed and acceleration so we could drive on the freeway and not be limited in where we go by speed or acceleration. With a top speed of 60 mph and a single charge range of 55 miles it meets these goals well." Courtesy Fred Weber.

The twenty-four 6V batteries are divided into two locations in the van; a front box mounted behind the driver's seat, and a rear box above the engine compartment. Shown here is the front box, which contains fifteen of the bulky batteries. In a pinch the front battery box can serve as an extra seat so the van can accommodate eight people. Courtesy Fred Weber.

containment of the batteries in the event of a crash.

The battery cages were constructed from 1.5" angle iron steel that was welded together at a local metal fabrication shop. Polypropylene boxes that line the cages were made from 0.25" material. Schedule 40 PVC pipe served as conduit both inside the van and underneath to connect the two battery boxes and to serve the rear motor and electronics. A large Curtis 1231-C motor controller was attached to a 0.25" thick sheet of aluminum that was suspended from a piece of angle iron bolted underneath the rear platform.

After he had finished constructing the battery boxes and cages, and after all of the conversion parts had arrived, Fred began a two day marathon, installing all of the EV gear with help from his brother-in-law. Late on the evening of the second day, they finished the wiring and immediately took the van for its inaugural drive around the block.

"There were no major problems or any debugging required on that first drive," says Fred. "Everything worked as it was supposed to. Since then, I have done a number of minor improvements. For instance, I added fans to the battery boxes with outside vents to remove any gas that might accumulate during charging."

Fred has also made a variety of custom parts and modifications to complete the project. "One of the personal touches I made is a change to the dashboard instrument panel. I wired the old Oil warning light to indicate that the key switch is on. Similarly, the Catalytic Converter light now indicates motor over-temp, and the Alternator light now tells me when I'm in reverse." Fred replaced the plastic lenses on these indicator lights with custom ones that have the proper symbols. He also mounted an ammeter in the dash where the original tachometer was located.

The family put over 1000 miles on the van in just the first six weeks and is very pleased with the conversion. Fred does say, however, that if he were to do the project over he would try to build in more range, possibly with larger capacity batteries. He also wishes he had searched a little longer for a better donor vehicle, as the current one had significant cosmetic surface rust that took a lot of time and money to remove.

"I would recommend to someone planning a conversion that they look for as perfect a donor vehicle as possible. Cleaning rust and restoring a car is not nearly as much fun as doing the electric conversion itself and actually took the most time."

When asked what is next for the van, Fred says that he is installing sound insulation and oak plywood paneling inside. The bus has also recently received a new professional paint job on the exterior. "And, because it's a VW van," says Fred, "there is one more thing that we have to do. My daughter is going to paint flowers and peace symbols all over the outside." After all, what better says "tree hugger" today than a Volkswagen Type 2 bus powered by electricity from the sun?

The front battery box shown with its cover in place and a steel angle-iron frame that clamps the container closed. Note the simple PVC pipe conduit used to carry the electrical cables underneath the van to the rear of the vehicle. Courtesy Fred Weber.

The battery boxes were constructed of heavy steel angle iron welded together at a local fabrication shop. Polypropylene material was used to form the walls, bottom, and lid of the boxes. The owner says that he got behind schedule during the conversion and did not paint the battery boxes before installing them. "I will likely regret that," he says. Courtesy Fred Weber.

The rear battery box installed in the vehicle. Note how more than half of the box extends down through the floor of the van into the engine compartment below. Courtesy Fred Weber.

Ventilation fans and PVC ducting were installed and are used to remove any gases that develop during recharging. This is an example of a simple solution to an important issue. Courtesy Fred Weber.

The owner modified the OEM dashboard indicator lights to work with the EV components. He also created new lens covers with the appropriate wording. Courtesy Fred Weber.

The van shown plugged into the grid through a standard 15A household electric outlet. Recharging takes approximately 12 hours on a 110V circuit if the batteries are fully discharged. Since the owner has a 7KW PV system on his roof, driving the van effectively has a zero carbon footprint. The owner reports that the van now serves as his family's primary vehicle for around-town errands. "We drive it about 40 miles during a typical day, taking kids to school, shopping, and so on. It works great for bringing home building supplies from the hardware store, and it carries lots of groceries. We're very happy with the conversion. It took a little while to master driving the clutch-electric motor system, but now I never think about it. I just get in the van and drive." Courtesy Fred Weber.

Project 9:

BMW Roadster

A well-executed 2000 BMW Z3 roadster that has been converted to electric power. The car has a 60-mile range and can top out at over 80 mph. Courtesy Tim Catellier.

Tim Catellier of Chandler, Arizona has wanted an electric vehicle for a long time. Back in 1997, he test drove General Motor's infamous EV1. The experience left him excited about the prospect of electric vehicles but also disappointed. The EV1 was a great car and economical to operate, but GM wasn't selling the vehicle, and the cost of leasing it was prohibitive. Tim simply couldn't justify the high monthly expense for what would effectively be a long-term rental.

Around the same time, Tim also test drove a Toyota RAV4 EV, but again he ran into a financial roadblock. At the time of the test ride, it was nearly impossible for a layperson off the street to walk into a Toyota dealership and purchase one of the vehicles. Toyota was focused on selling the mini electric SUVs to fleet owners and businesses that wanted to purchase ten or more RAV4s at a time. Prospective buyers of single vehicles were left wanting.

After years of waiting for Detroit, Japan, or Europe to step up and build an affordable performance-oriented EV that an enthusiast could actually buy, Tim came to a realization: If he wanted a well-engineered and affordable EV in his lifetime, he would have to build it himself.

"The first step was learning everything I could about EVs," Tim says. "Then I faced the question of choosing a vehicle to convert." He ended up selecting a 2000 BMW Z3 convertible because "it would be fun to drive, it looked nice, and I wanted a car that I would want to drive for years and years to come." Tim goes on to add, "People often convert older cars because they are inexpensive. That's fine, but when you're finished with the conversion, you can have an EV that may be, frankly, past its prime. Older conversion vehicles can be done well, but this wasn't for me. I was only really interested in newer vehicles for my conversion project."

Tim also wanted high performance and exceptional reliability. "I wouldn't call my EV a 'no compromises' build, but the truth is I didn't have to sacrifice very much." Tim chose the best components he could afford and installed them with a serious attention to detail. The idea was that if he purchased quality up front, he wouldn't have to replace anything in the near future and the car would serve reliably for a long time. Says Tim, "The basic philosophy was to purchase a newer vehicle, and then convert it using the best available parts in the best possible manner."

After locating and purchasing the low-mileage BMW Z3 shown here, the first order of business was to remove the old internal combustion engine and supporting equipment from the vehicle. This turned out to be an area where having a newer, more popular vehicle had an advantage over converting an older car or truck. "I found a local shop that wanted the ICE," says Tim. "I drove the car to their garage and they pulled the engine. They even paid me for it, as it had low mileage and was in excellent shape."

With an empty engine bay in the car open and waiting for electric power, the next step for Tim was to begin shopping for components. The entire build, from start to finish, took a little over one year to carry out. The first five months, however, were spent mostly just shopping and waiting for parts to arrive. The actual hands-on conversion in Tim's garage took only seven months.

SPECIFICATIONS

Vehicle: 2000 BMW Z3 Roadster
Original Engine: 2.5-liter 6-cylinder gasoline internal combustion
Motor: NegGain Motors WarP-11 11" series-wound DC motor
Motor Adapter: Electro Automotive
Controller: Café Electric Zilla Z1K-HV controller, Hairball interface, Zilla hall-effect pedal assembly
Batteries: 48 Sky Energy 120AH, 3.30V, lithium-ion
System Voltage: 160V
Charger: Manzanita Micro PFC-20 w/ Buck enhancement onboard charger
DC-to-DC System: Two Iota DLS-55 DC-to-DC converters, lead-acid gel-type backup battery
Wiring: 2/0 ultra flexible welding cable
Switches & Contactors: Tyco Kilovac LEV200
Fuses/Breakers: Ferraz/Shawmut A30QS500-4
Instrumentation & Gauges: Xantrex Link 10 e-meter
Safety: Main power disconnect switch in engine bay
Drivetrain Modifications: MR2 electric power steering pump; power brake vacuum pump and reservoir
Interior: 1500 watt 12V ceramic heater
Charging Time: 4 hours from a 240V line
Range: Approximately 60 miles
Top Speed: 80+ mph
Acknowledgments: "I would like to thank Tim Millward for all the welding, my father, Bill, for all the advice he offered, and my wife, Mary, for all the patience she showed."

The motor selected to power the Z3 was an 11" DC series-wound WarP 11 motor from NetGain Motors. Originally, a 9" motor was considered, but Tim wanted fast acceleration and the ability to cruise comfortably at freeway speeds without overtaxing the motor. Discussions with NetGain engineers convinced him to spend the extra money and purchase the larger motor from the start.

Tim took the same bigger-is-better approach when selecting his controller, which is a Café Electric Zilla Z1K-HV unit designed for battery bank voltages between 72 and 300 volts, and can regulate up to 1000 amps of motor current. A standard Zilla Hairball interface board was also incorporated in the circuitry for all input, output and programming interface duties. Both of these items, along with the main power contactors, relays, and fuses were located in a small sealed sub-compartment in the engine bay that BMW originally used for the OEM vehicle engine control unit and electronics. "This was the obvious location in the car to mount the controller and supporting electronics," says Tim. "When the hood is lowered, the sub-compartment is sealed off from water and the elements, which keeps the electronics dry and safe."

Tim didn't skimp on batteries either, selecting a set of 48 Sky Energy 120AH, 3.30-volt Lithium-Ion units. The nominal Z3 system voltage is 160 volts, split into two separate banks. Tim placed 31 of the batteries up front under the engine hood and the other 17 in the trunk. The total energy capacity of the batteries is approximately 19kW-hours, which, at 320 Wh/mile provides a theoretical range for the roadster of nearly 60 miles.

Mounting the front 31 batteries was relatively straightforward. He built three trays out of angle iron that he mounted to the chassis. Then he bolted the batteries into three plastic boxes and in turn bolted the boxes to the trays. More creativity, however, was required at the rear of the car for the 17 trunk-mounted batteries. "I cut the floor out of the trunk to install the batteries," says Tim. "I also built a steel frame that reinforced the hole I cut. The downside is I had to give up carrying a spare tire with this configuration, but I was still left with a relatively large, usable trunk. This was important to me; I wanted to be able to carry groceries and other small items. I figure if I have a flat, I will just have to call AAA. For me, this is a small price to pay for owning the car."

Tim also cut out part of the forward bulkhead in the trunk to mount a Manzanita Micro onboard charger in the space where the original OEM gas tank resided. Two Iota DC-to-DC converters were also located in this space. Tim says he needed two converters because of the amperage requirements of the electric power steering system, which can draw up to 80 amps. Each Iota DC-to-DC converter is capable of providing 45 amps, so it was a simple matter of wiring the two units in parallel, which provides a 90 amp capability. A small lead-acid gel-type battery was mounted in the corner of the trunk to provide emergency backup to the converters.

The power steering unit used was an electric Toyota MR2 pump and reservoir. Tim installed proximity switches on the steering column, which activate the pump whenever the steering wheel is turned to the right or left of center. An off-delay relay timer circuit was also incorporated into the design. This circuit keeps the pump switched on for at least ten seconds whenever it is activated, thereby keeping the unit from cycling on and off when the car is driven in parking lots. Once the steering has been straightened out for at least ten seconds, the pump switches off, keeping energy usage to a minimum.

Inside the cabin, the changes from ICE-power to electric power are more subtle than under hood. Tim installed a small Xantrex Link 10 digital e-meter in the center of the dash where the OEM

The motor selected for the conversion was an 11" NetGain WarP 11-type series-wound DC motor. Here the motor is shown installed in the engine bay. Note the Electro Automotive adapter plate that mates the motor to the stock BMW five-speed transmission. Courtesy Tim Catellier.

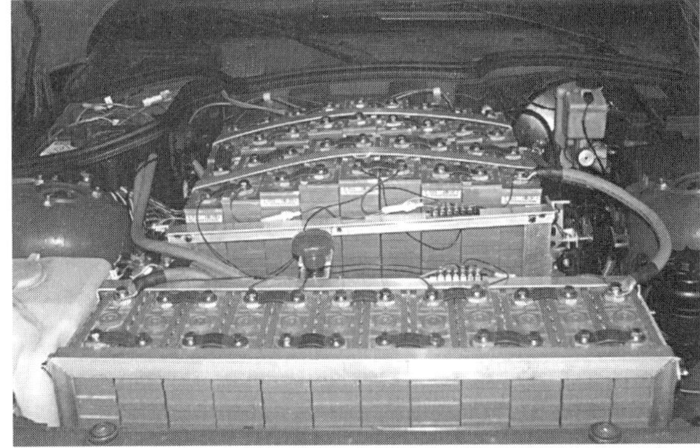

The owner installed 31 of the total 48 Sky Energy 120AH, 3.30 Volt Lithium-Ion batteries up front in the engine bay, directly over the motor assembly. The remaining 17 batteries were placed at the rear of the vehicle in the trunk. The nominal system voltage is 160 volts, and the total energy content of the 48 batteries is approximately 19kW-hours. Courtesy Tim Catellier.

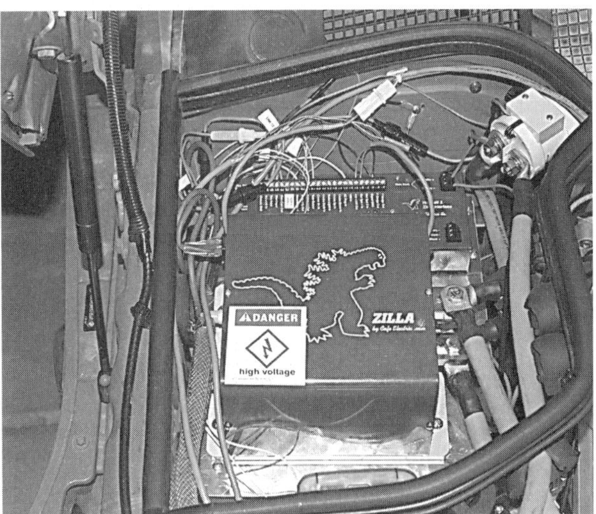

A Café Electric Zilla Z1K-HV controller and Hairball interface unit were used to control the motor. Note the large main contactor in the upper right part of the photograph. A number of relays and fuses are located underneath the Zilla. Courtesy Tim Catellier.

clock once resided. The Link 10 displays the current pack voltage, instantaneous current draw, and a running total of the kW-hours that have been pulled from the battery pack since the time it was last recharged.

The original liquid-type heater core was also replaced with an electric ceramic type. These units can get very hot if air is not blown over the top of them, so Tim wired the heater control in series with the fan motor switch. This does not allow the heater to receive electricity if the fan isn't also on at the same time.

Tim reports that building the car was a very rewarding experience. "When I first drove the Z3 out of the garage, it was one of the most satisfying things I've ever felt. I'm not an electrical engineer, so I think my success goes to show that if you're willing to read up on EVs and learn a few things and then apply that knowledge, anyone can build a high-quality conversion. If I can do this, anyone can." In other words, if you can't wait for Detroit, Japan, or Europe to create an affordable high-performance EV, you can always just build your own.

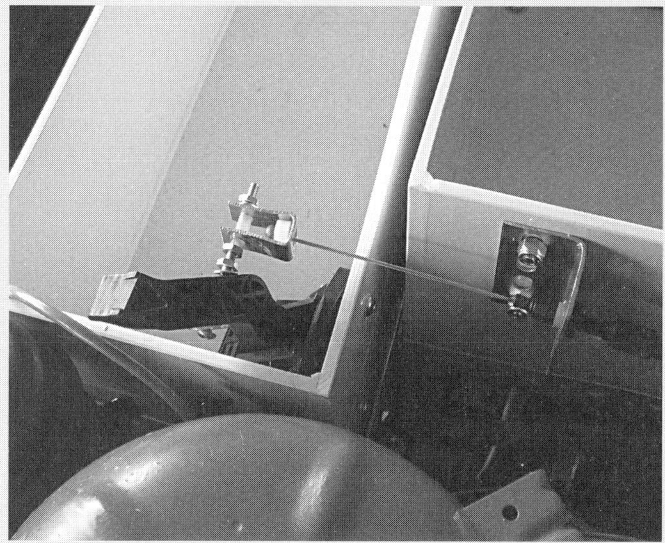

The original throttle cable was used to connect the OEM BMW throttle pedal inside the cabin to this cut-down Zilla hall-effect pedal that is mounted in the engine bay. Mounting it this way eliminated the need to modify the OEM BMW pedal assembly. Unique and effective. Courtesy Tim Catellier.

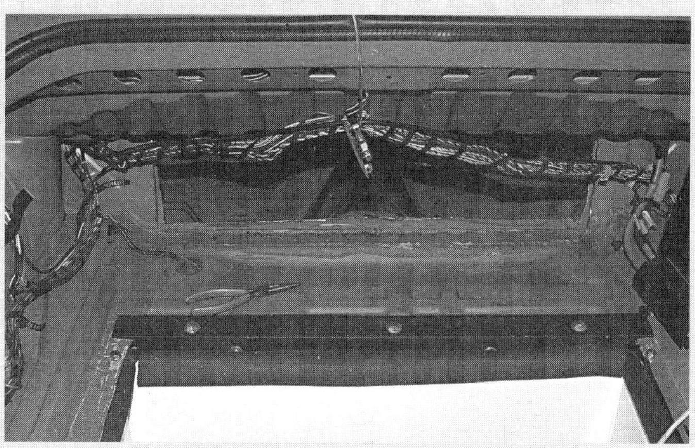

The rear trunk area of the Z3 was extensively modified for mounting the batteries, charger, and DC-to-DC converters. This photo shows how the bottom of the steel trunk floor was cut out and replaced with a reinforced battery tray. Also note the large square hole cut in the forward bulkhead of the trunk. Courtesy Tim Catellier.

A view showing the batteries mounted in the floor of the trunk. Also visible in this photo are the two Iota DC-to-DC converters and the Manzanita Micro onboard charger recessed into the front bulkhead wall of the trunk. The owner does not use a battery management system (BMS) with the lithium-ion batteries. "Some people think this is crazy, but so far it's been unnecessary. After charging, the batteries seem always to be in balance, day or night, whenever I test them. The secret seems to be in keeping the batteries in their sweet spot. Don't overcharge them, don't over drain them, and everything works out fine." Courtesy Tim Catellier.

A view of the central dash inside the Z3 cabin. A Xantrex Link 10 e-meter was placed in the central hole where the original BMW clock resided. Immediately to the left of the Link 10 is a rocker switch that operates the new electric heater. This switch is wired in series with the fan control switch (upper right), to ensure that the heater is operated only when the fan is blowing, thereby reducing the risk of overheating and fire. Courtesy Tim Catellier.

APPENDICES

Appendix A:
Electricity

"We believe that electricity exists, because the electric company keeps sending us bills for it, but we cannot figure out how it travels inside those wires." —Dave Barry

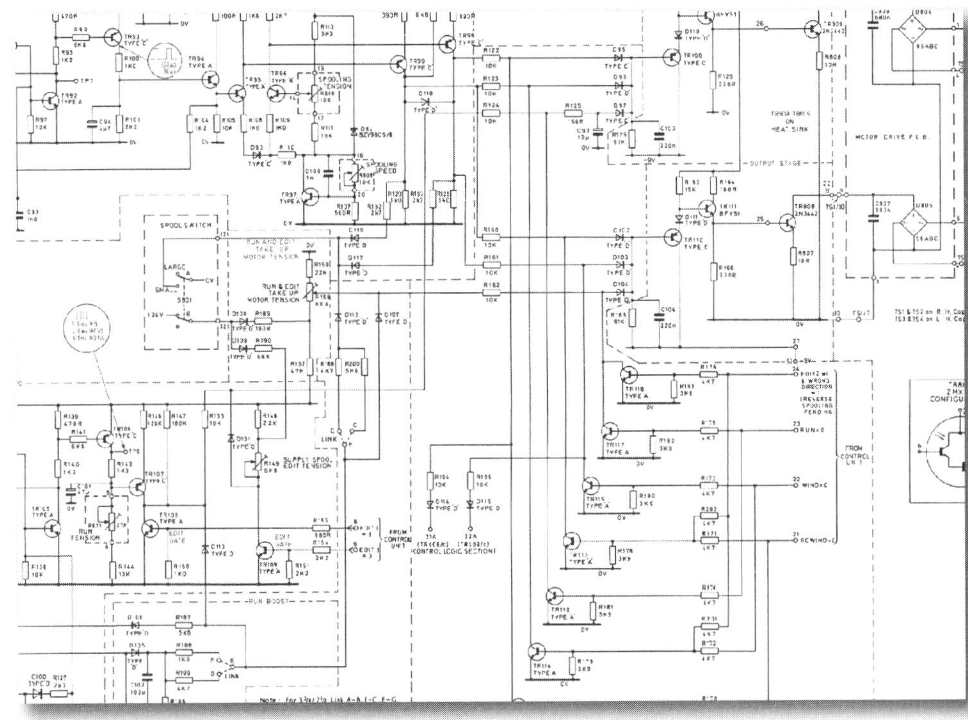

It may look complex, but even a sophisticated controller circuit like this one can be broken down and understood using simple node analysis and Ohm's law.

This is a book about converting your car, truck, motorcycle, or bicycle to electric power. While it's possible to jump right into the details of selecting components and assembling an electric vehicle (EV) without understanding basic electricity, it will help significantly if we first understand some simple concepts about electricity and electromagnetism. How electricity is created, stored, and put to use in simple electrical circuits is the focus of this section. Don't worry; you won't need a Masters degree in engineering to grasp these concepts. In fact, it's surprisingly easy to understand the fundamentals of what electricity is and how it works in simple DC circuits. Comprehending the basics is important to understanding the larger concepts of EVs—namely, how we can use a battery to make an electric motor spin, and then how that motor in turn can propel a car down a road or highway. Let's start with the concept of electricity itself.

Simple Electrical Circuits

There are many definitions and categories of electricity, ranging from the static electricity we experience when rubbing our feet across a carpet and then touching a door knob, to lightning strikes between clouds and the ground during a storm, to electromagnetic fields and magnets. While those are all useful concepts (and we in fact cover some of them throughout the book), we need to start with something more basic: the simple flow of electricity as a direct current (DC). To do this, let's look at a simple DC circuit that consists of a battery, some wire, a switch, and an electrical device, such as a light bulb. When we put these particular pieces together, we can create a simple DC circuit, like that found in a flashlight. Let's see how this works.

The typical battery found in a flashlight has a positive terminal and a negative terminal. Inside the battery, a chemical reaction produces a flow of electrons, which collect on the negative terminal of the battery. With a battery just sitting on a shelf, the electrons produced mostly just sit there on its terminal, not doing much of anything because they do not have anywhere to go. Enter the wire.

In our simple flashlight circuit, the wire acts as a pathway, or conductor for the electrons. If we were to connect the positive and negative terminals of the battery directly together with a piece of wire, the electrons would immediately begin flowing through the wire. (This is actually a very bad thing to do, as the battery will quickly wear out and, more important, a dangerous overload of the circuit will likely occur; i.e., don't try this at home.) The flow of electrons precipitates a charge that moves through the wire in the opposite direction as the electrons. Technically speaking, the rate at which the charge moves is called the electrical current, but most textbooks and engineers think of current as simply the flow rate of the electrons through the wire.

In much the same manner as water can be made to flow through a pipe when subjected to a pressure at one end, electrons can be made to move through the wire like this when a voltage is applied. The battery supplies the voltage (which is the difference in charge potential between the two sides of the battery). The battery also provides the supply of electrons. The rate of flow of electrons through the wire is called an electrical current, and it takes place via electrical

Origins

The effects of electricity have been observed since ancient times when it was noticed that amber would attract small bits of straw and other light materials when rubbed vigorously. In fact, the word *electricity* is derived from the Greek word for amber: *elektron*.

conduction. While we could express electrical current in units of electrons per second, the number is so large that it is unwieldy and, frankly, unhelpful. Instead, scientists and engineers use a term called the *ampere*, or *amp* for short. (It's not very important that you know this, but one ampere is the equivalent of 6.24E+18 electrons flowing per second through a circuit. Said another way, one amp is equal to 6.24 billion billion electrons per second!)

Now, instead of simply connecting the wire in our circuit from one side of the battery to the other, we insert a "load" in the circuit. In our case, the load is the light bulb, but it could also be any number of other items, such as a motor, a resistor, a sensor, or other electrical device. In our circuit, the electrical current is made to flow through an incandescent light bulb on its way from one side of the battery to the other. Incandescent light bulbs are comprised of a thin filament of wire (often made of tungsten) suspended in a bulb that has either an inert gas or a vacuum inside. The wire itself is very small in diameter and consequently has a relatively high resistance to the flow of electricity. This results in heat as electrons are forced to flow through the tiny wire. A light bulb subjected to a high flow of current gets hot enough to cause this wire to glow, which is the source of light that allows a bulb to illuminate a dark room.

Adding a switch in our simple electrical circuit is the final thing we need to do to make the circuit more user-friendly. The switch effectively acts as an easy means of breaking, or disconnecting, the wire when switched off. Turn the switch on (i.e., complete the circuit) and current is allowed to flow from one battery terminal through the light bulb, and back to the battery. Turn the switch off, and the flow of current is stopped and the light turns off.

Ohm's Law

In the simple flashlight circuit example of the preceding section, we introduced three of the most important principles of electricity: voltage, current, and resistance. There are other important concepts (such as capacitance and inductance), but these big

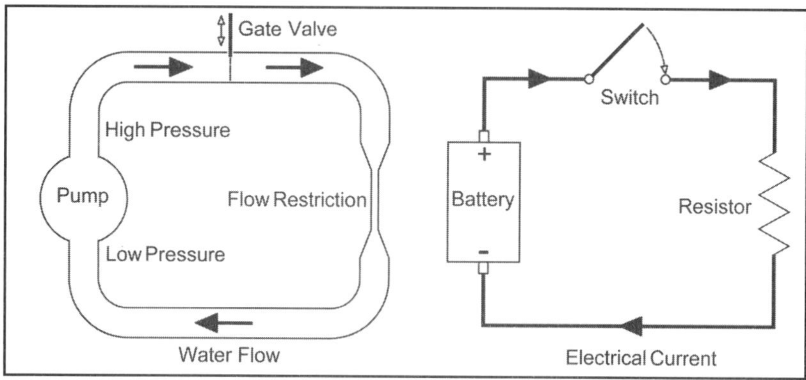

A simple way to understand electricity is to compare a simple DC electrical circuit to an equally simple water pump circuit, as shown here. The battery has a voltage difference across its terminals, in much the same manner as a pump has low and high pressure sides. This potential difference causes the flow of electrons in the wire, just as the pressure difference causes the fluid to move through the pipes. A switch can make or break the electrical circuit, just as a valve can allow or stop the flow of the fluid. If the pipe is long, or rough inside, or necks down to a small cross section, it resists the flow of the fluid. A resistor operates in a similar fashion.

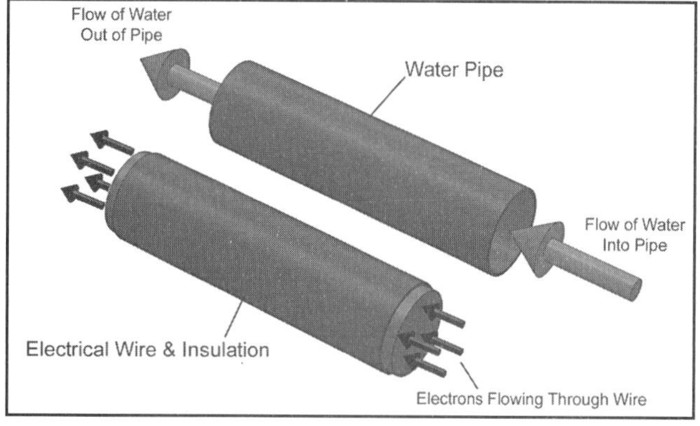

Electrical wire is like pipe in a water circuit. The larger the diameter wire, the lower its resistance to flow. Other factors that affect the wires resistance include the material it's made of, physically how long the wire is, and the operating temperature. Similarly, the flow of water in a pipe can be affected by how long the pipe is, how rough the inside surfaces of the pipe are, and what the temperature of the fluid is.

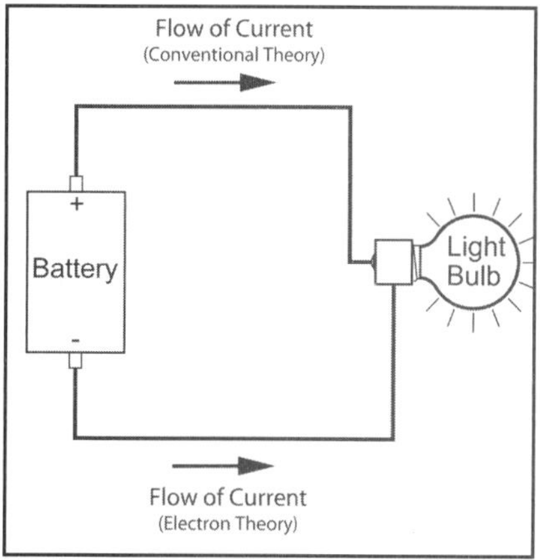

It may seem strange, but scientists and engineers can't agree on which direction current actually flows in a circuit. Consequently there are two theories used to describe current flow. The "conventional" theory, which is commonly used in automotive engineering, states that current flows from the positive (+) terminal of a battery to the negative (-) terminal. In other words, current flows from an area of high voltage potential to an area of low voltage potential. The "electron" theory, which is commonly used in electronic design, says that current flows from the negative terminal to the positive. Confusing? Yes. Important? Not really. The absolute direction of current flow in a circuit does not affect the three measurable units of electricity we care about the most: voltage, current, and resistance.

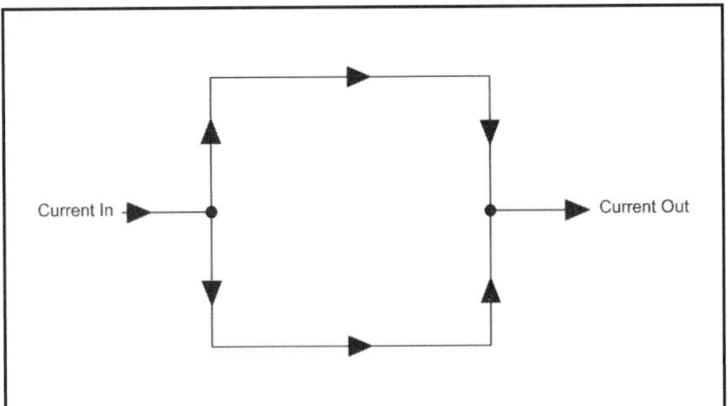

Just like water behaves in a series of pipes connected together, electrical current can branch apart and be rejoined. The current that enters a node will always equal the current that leaves the node. In the example shown here, electrical current enters from the left and then splits into two branches. The total current leaving the node will equal the amount entering. Similarly, on the right side of the circuit, the two branch lines come back together in another node. The currents combine at this point and exit to the right. Depending on the resistance of each branch, more or less current may travel in the upper or lower lines. Electricity, like water, will take the path of least resistance in a circuit. If the upper branch has higher resistance than the lower branch, proportionally more electricity will flow through the lower branch.

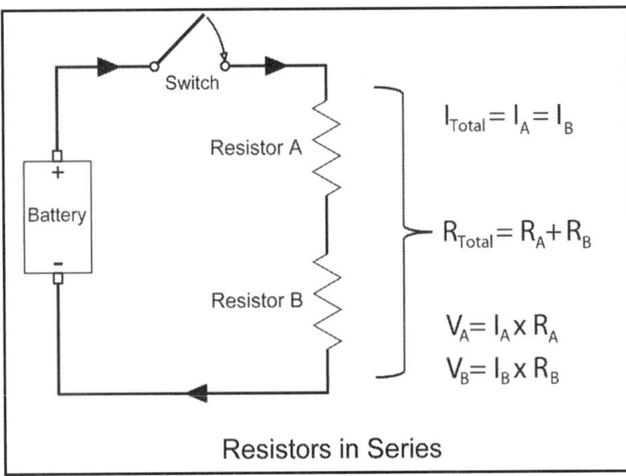

Resistors in Series

A series circuit is a circuit in which resistors are arranged in a chain, so the current has only one path to take. The current is the same through each resistor. The total resistance of the circuit is found by simply adding up the resistance values of the individual resistors: R = R1 + R2 + R3 + ... From Ohm's law, we can then calculate the voltage across each resistor, because we know its resistance and the current flowing through it: V = I x R.

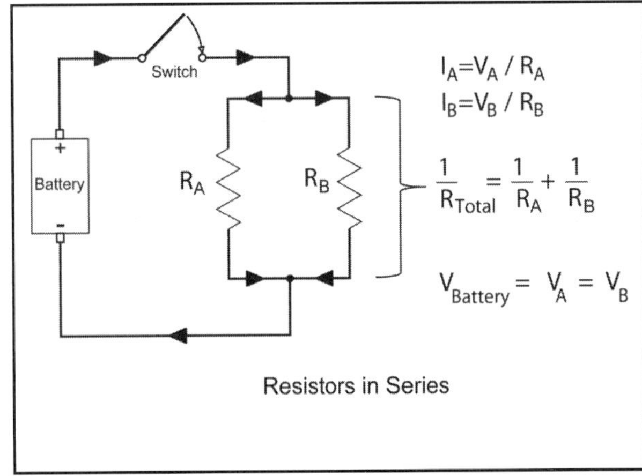

Resistors in Series

A parallel circuit is a circuit in which the resistors are arranged in side-by-side parallel branches, or paths. The current in a parallel circuit divides up, with some flowing through each parallel branch and then re-combining when the branches meet again. The voltage across each resistor in parallel is the same. The total resistance of a set of resistors in parallel, however, is not the same. It is found by adding up the reciprocals of the resistance values, and then taking the reciprocal of the total: 1 ÷ R = 1 ÷ R1 + 1 ÷ R2 + 1 ÷ R3 +... The current through each branch can be found by using Ohm's law: I = V ÷ R.

Basic Electricity Terms

Voltage is the term used to describe electrical potential differences in an electric circuit. It is usually denoted by the letter V in equations and is measured in units of Volts. The water pressure at the spigot end of a garden hose is a good analogy to electrical voltage in a circuit.

Current is the term used to describe the flow of electric charge in an electric circuit. It is usually denoted by the letter I in equations, and is measured in units of Amperes. The flow rate of water in a garden hose is a good analogy to electrical current in a circuit.

Resistance is the term used to describe the opposition to the passage of a current in an electrical circuit. The restriction caused by the hose itself or your thumb on the end of the hose is an analogy to electrical resistance in a circuit.

The *watt* is the electrical unit that defines the rate at which work is done, or energy is spent. Mathematically, we can say that watts = volts x amperes. The larger the current and/or the voltage, the faster energy can be converted into useful work.

three are the fundamental building blocks of most electrical systems, particularly those used in electric vehicles.

Voltage is typically measured in volts, and is often denoted with the letter V in equations and circuit diagrams. Similarly, current is measured in Amperes, and is, strangely, not denoted with the letter A, but instead is written with the letter I in equations. Resistance is measured in units of Ohms, and is denoted by the letter R.

In the early nineteenth century, a German physicist named Georg Ohm discovered that the electrical current flow through a conductive wire was directly proportional to the voltage applied to the wire and inversely proportional to the resistance of the wire. Written another way:

I = V/R

This is known as "Ohm's law of electricity."

So, why is this important? The answer lies in the equation itself. Voltage, current, and resistance all work in proportion to each other. In other words, if we know two of the three values in an electrical circuit, we can use the formula to predict what the third value will be. For example, if you know that the voltage of a battery is, say, 12 volts, and you know that the resistance of a motor wired into the circuit is 24 ohms, you could easily determine that 12 ÷ 24=0.5 amps of current would flow through the motor when it is in operation.

Using Ohm's law, we can also make changes to a circuit and accurately predict the result. If we add resistance, for instance, the current flow drops. If we increase voltage, the current goes up. We can also use the equation to select components and wiring, determine the amount of voltage needed to operate and electrical device, and so on. Ohm's law is at the heart of nearly all electrical system design choices.

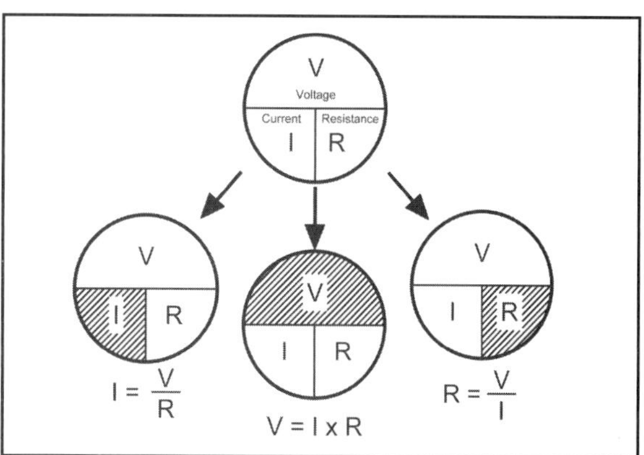

Ohm's law states that the current (I) through a conductor between two points is directly proportional to the voltage (V) difference across those points and inversely proportional to the resistance (R) between them. This means that if we know any two of these three things (i.e., voltage, resistance, and/or current) in a circuit, we can solve for the missing third item. In this image, the horizontal line in the circle represents division, while the short vertical line represents multiplication. If you don't know what the current is, for example, you can calculate it by dividing voltage by resistance. Similarly, voltage = current multiplied by resistance, and resistance = voltage divided by current.

DC vs. AC Circuits

The simple circuits we have discussed thus far in this appendix are examples of DC, or direct current, electrical devices. In DC circuits, the current (i.e., the movement of the charge via the electrons) occurs in one direction in the wires; namely, from the negative terminal of the battery, through the light bulb, and on back to the positive terminal of the battery. The current always flows in one direction in a DC circuit and does not reverse. This type of electrical circuit is used to operate many simple electro-mechanical devices, including most of the equipment used in a modern car (e.g., lights, turn signals, horn, power windows, etc.). It is also the most common type of circuit used to operate the motors that power homebuilt electric vehicles.

In contrast, many other types of electrical circuits utilize alternating current, or AC. For example, most of the electric and electronic devices used in a home or business are operated on AC electrical power. The garbage disposal under a sink, the alarm clock beside a bed, the refrigerator in a kitchen—all of these operate on AC power.

In an AC circuit, the electrical current periodically (and rapidly) reverses direction; that is, the charge (via movement of the electrons) moves in one direction for a period of time, and then reverses and flows in the opposite direction. This reversal usually follows a sine wave shape. In homes in the U.S., the current reverses sixty times per second. This reversal frequency is usually denoted in units of Hertz. In the U.S., the frequency is therefore 60 Hz. In Europe, AC power to homes and businesses is typically 50 Hz, or fifty times per second.

Neither AC nor DC power is better, per se; both have certain advantages and disadvantages. AC power, for instance, allows relatively efficient increases (or decreases) in voltage via a device called a transformer. For example, the electric power that is delivered in large overhead power lines in a city is AC at very high voltages. These high voltage lines are used because they suffer lower losses when transmitted over long distances. Once delivered to a home or business, however, lower voltage is required. Local transformers step down the high voltage to 120 volts for use in homes.

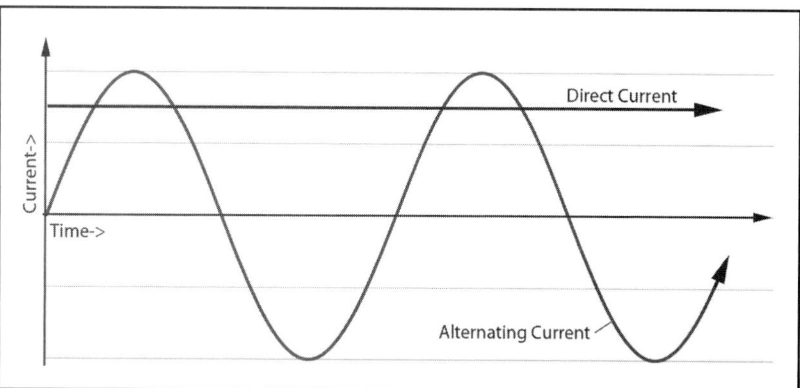

In AC circuits, the direction of the current changes with time, often in the form of a sine curve, as shown here. In the U.S., typical AC residential electrical power is 110–120 volts and it switches direction at 60 Hertz, or 60 times per second. DC current does not change direction with time.

Due to their simplicity and lower cost, most DIY-type electric vehicles are built using DC motors. Most commercial EVs, however, operate using AC motors and controllers, which allows for more advanced functions such as regenerative braking.

It is possible to convert AC to DC via a simple power supply. It's also possible to convert DC to AC, but it requires more sophisticated electronics, such as an inverter.

Appendix B:

Torque, Work, Energy, and Power

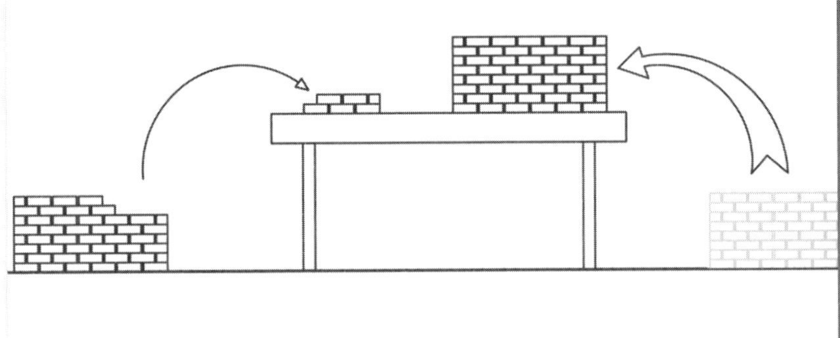

Lifting 100 lb of bricks from the floor up onto a three-foot-high table results in 100 lb x 3 ft = 300 lb-ft of work being performed. If a small child performs this task by lifting each onto the table, he or she expends the same energy as a strong adult who lifts all of the bricks onto the table at one time. Physicists say that both the child and the adult have performed the same 300 lb-ft of work. Now, power is the rate at which work is performed. Because the adult who lifts the entire brick stack takes much less time than the child's brick-by-brick effort, the adult generates far more instantaneous power than the child. Power is work divided by time, and it can be measured in watts, horsepower, or lb-ft per second.

"To perform work, you must expend energy."
—Physics 101 textbook

Okay, so an electric motor converts chemical energy into useful work performed by an electric motor. And that motor generates a certain amount of torque and power. Fine, but what do these terms actually mean? And why is it important to understand them?

To begin, let's define the word *power*. Physicists and engineers like to say that power is equal to the rate at which work is performed by a device, such as an engine. They go on to define *work* as what happens when a force physically moves an object a certain distance. If we want to move, say, a 100-lb stack of bricks from the floor up onto a table that is three feet high, we need to perform 100 lb x 3 ft = 300 lb-ft of work. It doesn't matter whether a small child performs this task by lifting each brick individually, or whether a strong adult does it in one single hernia-inducing lift; both the child and the adult will perform the same 300 lb-ft of work if all the bricks are lifted onto the table.

But there is clearly a difference in the amount of power expended in each case. Because the adult can perform the task much more quickly than the child, we say that the adult is capable of performing more work per unit time than the child. Put another way, the adult can generate more power than the child when performing the same task. If, say, the adult takes five seconds to lift all the bricks, he generates 300 lb-ft per 5 seconds = 300 ÷ 5 = 60 lb-ft/seconds of power, which is equal to about 0.1 horsepower. If the child takes two minutes, or 120 seconds, to perform the same task, brick by brick, he produces only 300 ÷ 120 = 2.5 lb-ft/seconds, or about 0.005 horsepower. As we can see from this example, power is a function of three things: force, distance, and time.

This is all well and good for bricks and straightline systems, but our electric motors don't operate via linear movements. We need to somehow take this notion of force, distance, and time, and apply it to a system that rotates. The trick to doing this is via the concept of torque.

Torque is the result of a linear force acting at a specific distance from a rotational center point of an object, causing that object to try to rotate about that center point. Torque is readily calculated by multiplying the linear force by the distance from the rotation center. If, for example, we push with 20 lb of force at the end of a one-foot-long wrench that is fitted to a bolt, we generate 20 lb x 1 ft = 20 lb-ft of torque that is trying to turn the bolt. In a sense, torque is the rotational equivalent of a force and, in fact, is often referred to as a *twisting force* by armchair physicists.

Now, it's important to note that we can create and apply a torque to the wrench, whether the bolt actually turns or not. If the bolt doesn't turn, no useful work is performed by the application of torque. If the bolt does turn, however, work is indeed performed. And just as straightline work is defined by the product of a linear force and the linear distance traveled by an object, rotational work is measured by the product of the applied torque and the resulting angular amount that the object physically turns.

So, now that we have torque and angular rotation, we can calculate *power* in a rotating system if we know the third variable: time. In a sense, we're now using torque, angle, and time instead of force, distance, and time to calculate power. If we know how much time it takes for the applied torque to cause the object to rotate an angular amount, we can simply divide it into the work and calculate the power developed. But before we do this calculation, let's simplify things a little bit:

Power = torque x angular rotation ÷ time

And noting that angular rotation ÷ time is really nothing more than rotational speed, we can rewrite our equation as:

Power = torque x rotational speed

We have to be a little careful to be consistent with the dimensional units when we apply this formula, especially when we want to know power in terms of horsepower, and we know rotation speed in units of rpm. Working through the math and simplifying things a bit gives us:

Power [lb-ft/sec] = torque [lb-ft] x rotation speed [rpm] x (2 x Pi) [radians/rotation] x (1 ÷ 60) [minute/second]

Or:

Power [lb-ft/sec] = 0.1047 x torque [lb-ft] x speed [rpm]

To change this into horsepower, we have to divide by a conversion factor of approximately 550 lb-ft/s per horsepower. If we do this and then rearrange the formula a bit more, we arrive at the classic equation:

Power [hp] = torque [lb-ft] x speed [rpm] ÷ 5252

This equation is a universal formula that links torque and motor speed to horsepower. It doesn't matter whether we're talking about four-cylinder Honda gasoline engines or an Advanced DC electric motor; the formula is valid for calculating power. It's also the reason that graphs of torque vs. rpm and horsepower vs. rpm, when plotted on the same scale in units of lb-ft and horsepower will always cross each other at 5252 rpm.

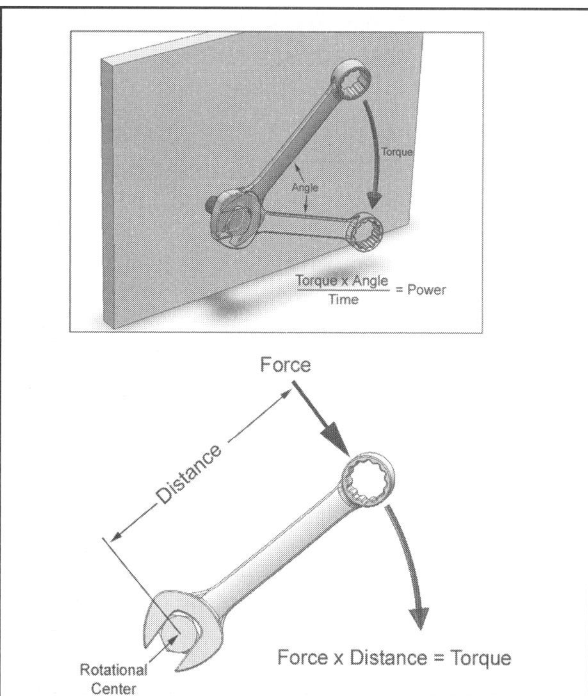

Power is nothing more than the rate at which work is performed by a mechanical system. In the case of a rotational device, such as a wrench turning a bolt, power is the applied torque multiplied by the angle that the wrench turns and divided by the time it takes to turn the wrench that distance. If we recognize that the angular distance divided by the time is nothing more than the rotational speed of the operation, we can rearrange the formula a bit and arrive at a standard equation: Power = torque x rpm ÷ 5252, where torque is measured in lb-ft.

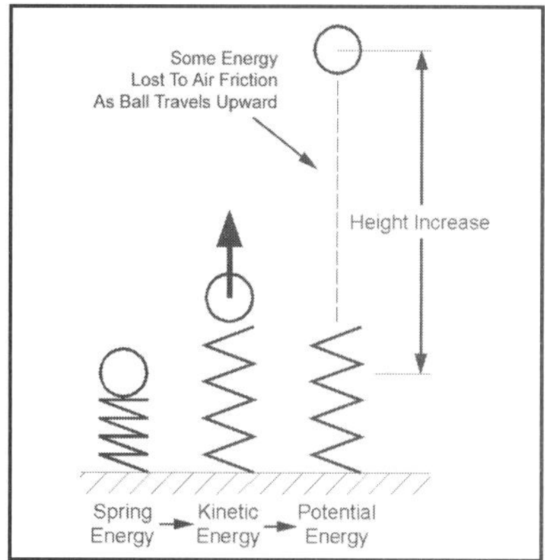

In physics, the term *energy* is the capacity of a physical system to do *work*. There are many forms of energy, including stored energy, such as the energy stored in a compressed spring. The energy of movement is called kinetic energy, and it depends upon the mass and speed of the object. Potential energy is an energy associated with raising a mass to height and is, in fact, a type of stored energy.

Batteries are an energy storage and conversion device. When charged, the internal chemical energy can be converted to electrical energy by applying a circuit to the two terminals. Similarly, an external source of electricity can be used to recharge the battery, essentially converting electrical energy back into chemical energy. Courtesy Deka Batteries.

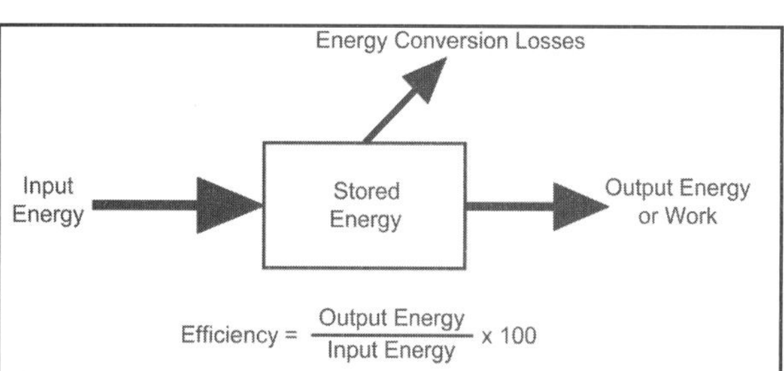

As we all learned in high school science class, energy can neither be created nor destroyed. It can only be stored, transferred to another location, or converted to a different type. We also learned that there are no 100% perfectly efficient devices that can convert energy; there are always some type of losses. Friction, heat, and other parasitic phenomena always rob us of some energy when we try to store or convert it. When we charge a battery, for instance, some of the input energy is converted to heating the electrolyte and plates. This heat is eventually "lost" through the battery case walls, never to be recovered again. Similarly, when the battery is discharged, some parasitic heat loss is incurred again. The efficiency of this process can be easily calculated by dividing the measured output energy by the input energy required for recharging.

Appendix C:

Licensing and Registering Your EV

Before beginning an EV conversion project, it's recommended that you check with your local motor vehicle department to understand the requirements of licensing and registering the vehicle. It is possible to register an EV in every State in the U.S., but some local DMVs make the task more difficult than others.

"If you want to know what life [was] like in the Soviet Union, just go to your local department of motor vehicles." —Billy Joel

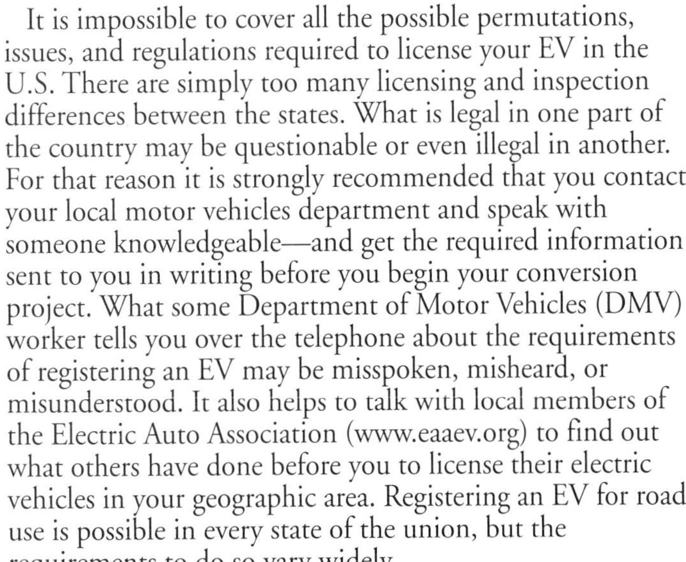

It is impossible to cover all the possible permutations, issues, and regulations required to license your EV in the U.S. There are simply too many licensing and inspection differences between the states. What is legal in one part of the country may be questionable or even illegal in another. For that reason it is strongly recommended that you contact your local motor vehicles department and speak with someone knowledgeable—and get the required information sent to you in writing before you begin your conversion project. What some Department of Motor Vehicles (DMV) worker tells you over the telephone about the requirements of registering an EV may be misspoken, misheard, or misunderstood. It also helps to talk with local members of the Electric Auto Association (www.eaaev.org) to find out what others have done before you to license their electric vehicles in your geographic area. Registering an EV for road use is possible in every state of the union, but the requirements to do so vary widely.

Okay, that said, there are some general guidelines to keep in mind when preparing to register an EV for road use. For instance, a full safety inspection prior to registration is required by many DMV offices. Everything from body and frame integrity to operating lights, hazards, turn-signals, and horn can be subject to inspection. Are the mirrors installed and functional? Does the windshield have cracks? Are the correct number and type of seat belts installed? What condition are the brakes in? How much play is in the steering system? Do the tires have adequate tread depth? Are there breaks, cracks, or severe rust on the frame or at the suspension attachment points? The list of items that are checked can be long and involved. Don't try to skimp on any of these items, either; your own safety can depend up on all of these things being up to the task of safely transporting you and your passengers at speed on the highway. In fact, some EVers hire an independent safety consultant to inspect their vehicles even if they're not subject to approval by their local DMV.

You also need to be aware of the local regulations concerning inspections and approvals if you reconstructed a non-operational "salvage" vehicle; some of these inspections can be much more involved and may even require you to produce receipts for the components you used during the build.

Another consideration is if the vehicle is custom built from components and parts. Some states make registration of this type of vehicle easy, others much harder. At the time this book was being written, California, for instance, with a population of 37 million people and over 30 million registered cars, trucks, and motorcycles, limited the number of custom built vehicles (like EVs) to just 500 per year. Registration spots are almost always gone on the very first available day of the year. Crazy, but true for a state that likes to think of itself as extremely progressive and on the leading-edge of environmental causes.

You also need to be aware that some vehicles may not qualify for registration even if they pass every safety check and inspection. For example, certain imported subcompact cars, off-road vehicles, and other cars and trucks that don't normally qualify for licensing will probably still be illegal, even if they're converted to electric power. (One exception to this may be registering the vehicle as a low-speed "neighborhood electric vehicle," or NEV.)

Finally, smog and emissions inspections may actually present a problem in trying to register an EV. Obviously, an electric vehicle has zero tailpipe emissions, but in the bureaucratic minds of some local governments, an EV may technically fail a smog test—not because it produces pollution, but because it no longer has any of the original anti-pollution equipment (e.g., catalytic converter, PCV valve) still installed. In other words, an EV may fail the visual inspection portion of a smog test. There are tricks to getting around this insanity, but it often involves getting a "smog referee" involved, who can recommend an exemption.

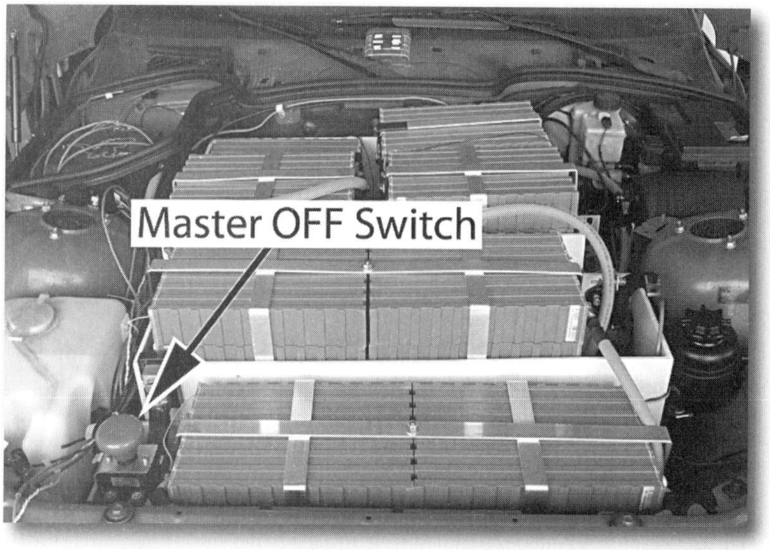

Master OFF Switch

A large red button located in a very convenient place under the hood allows the owner of this electric-powered BMW to easily isolate the battery pack from the rest of the system. Making this type of breaker switch readily accessible means that it will more likely get used whenever the owner performs maintenance in and around the engine bay. Courtesy Tim Catellier.

Appendix D: EV Safety

"There are two good rules for a longer life: assume the gun is loaded, and the electric wire is hot. There will be fewer surprises that way."
—Stuart Perkins, EV enthusiast

So you've read this book cover-to-cover, plus everything else you can get your hands on about EVs. You've scoured the internet, talked to experts, and wrangled as many test rides you can from fellow EV owners. You know what all the parts of an EV are, how they work together, and where to buy them. You are primed and ready to jump in and start a conversion project. What could be simpler? Some batteries, a motor, a controller, and some wiring, right?

Wrong.

Building an EV may be simple in theory, but in practice it is a little more difficult than simply wiring up a motor and batteries and driving happily off down the road. More important, if you do things even a little incorrectly, you can end up with an unsafe vehicle that could self-destruct or, worse, hurt you or your passengers. Like any large and powerful machine, an electric vehicle has to be treated with respect—or bad things can and will happen.

In this short—but very important—section, we will take a look at some of the important safety considerations that need to be addressed before you purchase any component or pick up the first wrench. We will begin with a look at designing a safe EV from the start, including some thoughts on chassis selection, and continuing through layout and configuration of the major components. We will then look at construction techniques, including workshop safety and household charging considerations. Included in this we will also review some commonsense ideas on testing and shaking-down your new EV.

Building, owning, and operating a modern EV can easily be done in a safe and prudent manner—provided you are armed with the right information and have a healthy respect for the electrical, chemical, and mechanical equipment that comprise the conversion project. Whenever you're in doubt about something, stop, learn, and understand before proceeding. Building an EV isn't difficult, but it does take a commitment to safety and caution. Let's start the discussion with planning a new EV conversion, beginning with chassis selection itself.

> Electric vehicles are safe if treated with respect. Gasoline-powered vehicles are dangerous, too, but we're all used to them and know that we turn off the engine and put out cigarettes before refueling, don't open hot radiators, touch exhaust systems, run the car in closed spaces, etc. EVs are not any more dangerous, but they present a different set of potential problems. Take the time to learn what they are and you will be much safer in the long run.

Vehicle Selection

Before you can answer questions like "Do I need a sedan or truck" or "What motor should I purchase" or "Do I want electric windows or not" when planning an EV conversion, you need to first understand a bit about basic structural soundness and capacity of the chassis you are considering for conversion. It doesn't matter if you're planning a 300-lb motorcycle or a 3000-lb truck—if the body, chassis, and suspension aren't up to the task of safely carrying the heavy weight of the batteries and other EV equipment, you are destined to have problems.

A good starting point to determine whether your vehicle choice can safely carry the loads of conversion lies in something called the manufacturer's gross vehicle weight rating, or GVWR, which is the maximum allowable total weight of the vehicle. This GVWR includes the weight of

All vehicles sold in the U.S. have a Gross Vehicle Weight Rating, or GVWR, sticker affixed somewhere to the body or frame, usually on the driver's side door frame or under the hood. The GVWR is the maximum allowable total weight of the vehicle, including the weight of the vehicle itself plus fuel, passengers, and cargo. This number can be used as a basis of determining the maximum allowable load of batteries a conversion candidate vehicle can safely carry. Also useful are the front (FRT) and rear (RR) Gross Axle Weight Rating (GAWR) ratings that are shown on the placard. The GAWR ratings are the maximum distributed weights that may be supported by each axle of a road vehicle.

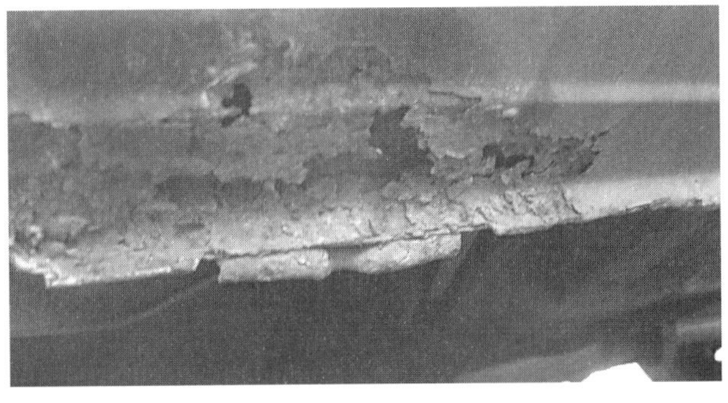

When looking at candidate vehicles for your conversion project, take the time to check for rust, corrosion, or collision damage. Cars like this should be passed over in lieu of examples in better condition. If you're in doubt, have a mechanic look over the car; most reputable shops will perform a thorough inspection for a nominal fee. Take your time when deciding on a conversion chassis. A few extra days or weeks spent shopping for the best platform that can be safely converted will be worth the peace of mind after all the hard conversion work is completed.

the vehicle itself plus fuel, passengers and cargo. The difference between gross vehicle weight and curb weight is the total passenger and cargo weight capacity of the vehicle. For example, a pickup truck with a curb weight of 3500 lb might have a cargo capacity of 1500 lb, meaning it can have a GVWR of 5000 lb when fully loaded.

Most U.S. cars have a placard or sticker with this information that is typically located either on the driver's side door or doorframe, or will be listed in the owner's manual. Once you have planned your project and selected batteries and components, you should make an estimate for how much weight will be added by the conversion. While it is true that when you convert a car you remove a number of heavy components (such as the engine, radiator, exhaust, and fuel tank), you will add back even more weight when you start adding EV components. Said another way: batteries are heavy.

> *"The exact nature of electricity—that it cannot be detected with the eyes, ears, or the nose, yet if it is touched, can kill—must be remembered at all times."* —L.W. Brittian, Mechanical-Electrical instructor

Very heavy, in fact. The typical 12-volt lead-acid battery can weigh between 50 and 85 lb, which means a standard 144V system will add between 600 and 1000 lb! Even if you remove 500 lb of ICE and supporting equipment from the vehicle, when you add back the batteries, electric motor, adapter plate, controller, and other equipment, the total vehicle weight can increase by a significant amount. The chassis that you select must be engineered to carry this load.

Also, the individual vehicle you're considering purchasing needs to be structurally sound. Vehicles with prior rust, corrosion, or collision damage should be passed over in lieu of better condition examples. Don't be shy about asking for "carfax"-type reports and other means of determining the history of the vehicle. Crawl underneath and look for damage and rust. If you're in doubt, have a mechanic look over the car; most reputable shops will perform a thorough inspection for a nominal fee.

Another thing to consider when looking at different types of cars is whether the brakes are up to the task of stopping the vehicle with the extra weight of conversion equipment. EVs can be very heavy, so power brakes are recommended. Four wheel disk brakes are best, too. Some cars can be fitted with even stronger brakes by swapping in larger rotors and calipers from different vehicles.

Take your time when deciding on a conversion chassis. A few extra days or weeks spent up front shopping for the best platform that can be safely converted will be worth the peace of mind after all the hard work is completed.

Component Selection

Once you've selected a safe vehicle to convert, the next step will be ensuring that all the components and equipment used in the conversion are also safe. In addition, implementing these items in a safe manner is vitally important.

For instance, when selecting a motor, make sure it's sized correctly. Too small of a motor (and/or too small of a controller) can result in the system being taxed too hard on the road. This in turn can lead to overheating and, possibly, a fire.

Ensure the motor has sufficient insulation, which is usually specified either by UL Class rating, or by

Wherever possible, install protective non-conductive boots or sleeves on exposed terminal connections. This minimizes the chance of an accidental short if a tool is dropped onto the leads. Courtesy EVSource.com.

The main chemical to be concerned with when dealing with an EV is the battery electrolyte. This is sulfuric acid (H_2SO_4) in lead-acid batteries. Use baking soda and water to neutralize any spilled acid. After neutralization, the leftover constituents are mostly harmless and may be cleaned up with soap and water. Wear proper protective clothing, including gloves and safety goggles while working with electrolyte.

allowable operating temperature. Some of the better EV motors commercially available are UL Class H, which is normally rated for 180 degrees Fahrenheit.

Elsewhere in this book we discuss fuses and circuit breakers. Do not skimp in this area. It is strongly recommended that you have at least two separate over-current protection devices in the main power feed from the battery to the motor controller. If possible, make these two devices separate types, such as a fuse and circuit breaker. The goal is to design a system such that no single failure of a component or safety device would be catastrophic.

Similarly, make sure you use appropriately sized fuses, fusible links, and/or circuit breakers in all other major and minor circuits. Small ancillary and auxiliary pieces of equipment and circuits can be overheated and catch fire just as easily as large systems.

Wire and cables should all be sized to carry the maximum system current at full loads. Use contactors and relays to switch on and off major electrical power circuits; do not use manual switches or circuit breakers inside the cabin. The goal is to keep all high voltage and high current circuits out of the cabin and away from the driver and passengers.

Use grommets and sleeves to ensure wires don't wear where they pass through holes in metal surfaces. Cars are subject to near-continuous vibrations when driven on even the smoothest of roads, and just a slight amount of contact between the sharp edges of a hole in a dashboard can quickly wear through the insulation layer of an electrical wire.

Except for the auxiliary circuits, never use the frame or chassis as a ground link for any of the EV circuits. The chassis is fine for lights and horns, but never for the main power systems. Charging problems, damage to the auxiliary components, over-heating, and dangerous shocks can all result if you try to save money on wiring by using the frame for return ground pathways.

When laying out components and equipment under the hood, try to locate critical connections and parts away from areas that can be splashed by water, road debris, and other hazards that might be kicked up into the underside of the vehicle.

Similarly, do not place contactors, switches, circuit breakers, or other items that can spark above or near your battery systems. We will discuss battery venting in a moment, but for now just remember that flooded lead-acid batteries can out-gas explosive fumes during charging.

Also, when wiring the vehicle, use terminal covers, heat shrink on connections, battery terminal boots, and other insulating devices to prevent accidental shorting when working on the system with metallic tools. Remember that batteries are always "on," whether the vehicle being driven or not.

Another consideration is the installation of a master kill or disable switch that is within easy reach of the driver. Having the ability to completely shut down the main power system (remotely, of course, via the use of a relay or contactor) can improve the overall safety of the vehicle.

Similarly, an inertia kill switch should be added to the system. Most EV suppliers carry these devices, which disable the electrical power system in the event of a crash by sensing high deceleration levels.

Finally, when planning your conversion, consider leaving as much of the original safety on the car in operating condition as possible. Seat belts, airbags, and the like can all be used in EVs. Have these items

Even if you are the only operator of the vehicle you build, take the time to label everything, including warnings about high voltages in battery boxes. The life you save might be your own. Courtesy Wayne Krauth.

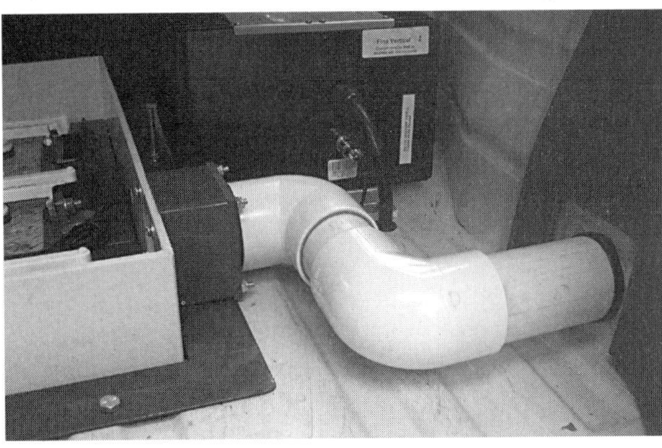

Lead-acid EV batteries can give off explosive gases when being charged. It's important to properly ventilate any battery boxes or containers. The owner of the electric VW van has installed small explosion-proof fan units on each battery box. Plastic PVC piping routes any gases from the box out of the vehicle. Courtesy Fred Weber.

> Remember: Batteries are always ON, regardless of whether the vehicle itself is powered on or off.

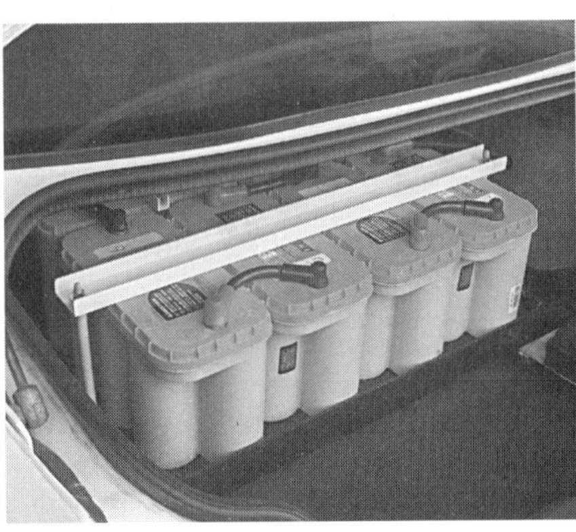

Batteries are heavy. Having a loose battery move around during vehicle operation can be unsafe. Having batteries come loose in an accident can be catastrophic. Take the time to construct solid frames and tie-down clamps for your batteries, as the owner of this electric BMW has clearly done. (Note that these batteries are AGM-types, which cannot spill acid in the event of a rupture; consequently, a sealed battery box is not necessarily required, as it might be for a flooded-type lead-acid battery bank.) Courtesy Rob Nicol.

checked for condition and operation by a qualified mechanic before driving the vehicle on the road.

Battery Safety

The means by which you mount and wire the batteries is a key factor in creating a safe electric vehicle. There are essentially four primary concerns that need to be addressed: weight, liquid containment, out-gassing issues, and stored energy concerns. Said another way, EV batteries are heavy. They're filled with corrosive chemicals. They can give off explosive gasses. And when wired together they can have enough volts and amps to kill a human. While this sounds dreadful, it's important to keep in mind that similar words of warning can be said about the heavy 10- or 20-gallon tank of highly flammable gasoline bolted underneath your average ICE-powered car or truck. A battery bank is indeed potentially dangerous, but it can be safely installed and used if the owner follows just a few common sense practices.

As we said, batteries are heavy. This means they need to be firmly fastened in such a way that they can't come loose during normal operation or in the event of accident. This means strong steel racks and hold-down clamps that secure the units in place.

Generally speaking, the lower the mounting position of the batteries, the better; i.e., the location of the vehicle's center of gravity can have a large effect on its handling. Mounting heavy batteries up high in the car can result in a vehicle that is unsafe to steer around corners.

In addition to steel racks and hold-down clamps, plastic boxes that line the racks are a good idea, as are insulated covers that can prevent accidental shorts when working around the batteries with steel tools. A properly designed battery box will also prevent any acid from entering the passenger compartment in the event of an accident. Battery acid is really only a problem on flooded lead-acid batteries; AGM and gel-type batteries have their acid immobilized and are essentially spill-proof. Battery acid is only moderately strong, but it can still cause burns to your skin if not washed off and should therefore be treated with respect. Wear safety glasses when working around open batteries to ensure that your eyes are protected from any accidental splashes.

More important than acid containment are out-gassing considerations, especially when charging is taking place. Flooded lead-acid batteries generate hydrogen gas when placed on a charger. A means of safely venting the batteries to the outside air needs to be factored into the design of the mounting boxes.

When designing the boxes and mounting

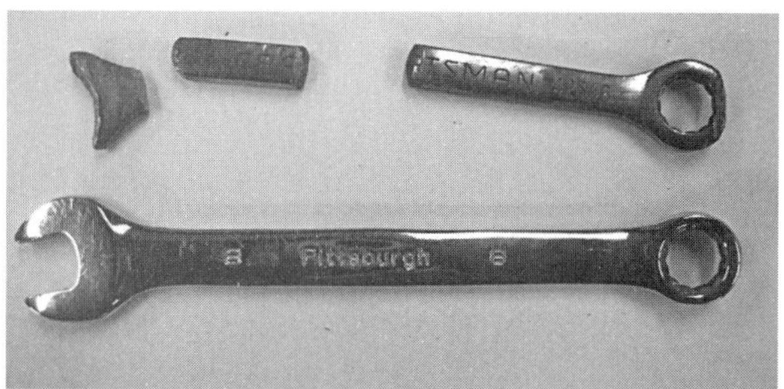

The owner of this 8mm wrench accidentally made contact across two points in his main power circuit that had a large voltage potential difference. In a matter of seconds, the wrench was glowing red and flames were rising out of the vehicle. An explosion followed shortly thereafter, leaving little of this wrench behind. As the owner said, "This only goes to prove that an EV builder with years of experience like me can still make mistakes. Learn from this and follow all safety precautions. I'm only alive now because I had insulated safety gloves on." Courtesy Tony Helmholdt.

locations, spend a little extra time to ensure that you will be able to safely service and water the batteries without having to reach too far or put yourself in an awkward and dangerous position. More than one person has leaned across an open battery box and shorted out the terminals with his or her belt buckle. To help minimize this type of possibility, you should consider installing insulated terminal covers on the battery posts.

You should also arrange the batteries in a manner that ensures the maximum practical separation between high voltage terminals. In other words, when laying out an array of batteries that are wired in series, try not to have the first and last batteries in a string physically located near each other. Accidently dropping a tool or other electrically conductive item onto the terminals could create a short circuit path that has the maximum voltage potential difference. This of course would result in the maximum short circuit current flow. To prevent this, when deciding on the battery locations spend a little extra time trying to minimize the number of locations where large voltage potential differences exist in close proximity.

You might also consider dividing your battery bank into two or more separate battery boxes. This helps minimize voltage potential differences between batteries (and boxes), plus it allows better weight distribution within the vehicle. For instance, half of the batteries can be located at the front of the car and half at the rear in the trunk, thereby keeping the fore-aft weight balance similar to the original ICE configuration of the vehicle. If you do separate your batteries into two or more separate

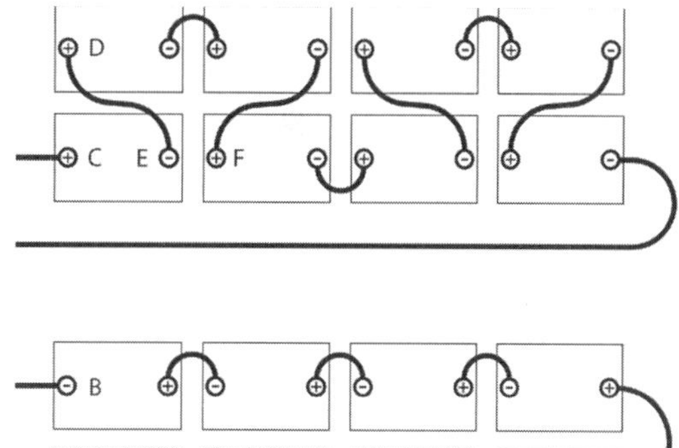

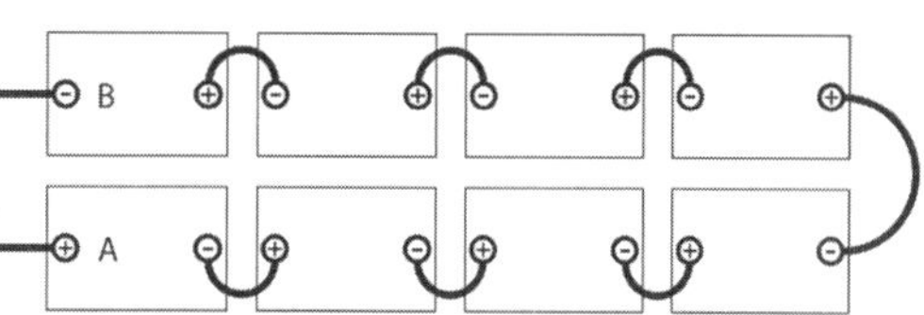

Shown here are two wiring layouts that connect eight 12-volt batteries in series for a total system voltage of 96 volts. The bottom circuit uses a number of short, direct connections between adjacent batteries. This simplifies the construction and wiring but also results in a very large voltage potential difference between the first and other batteries in the series. A wrench accidently dropped across terminals A and B will result in a coupling of 96 volts. In comparison, a wrench dropped across terminal C and D in the upper circuit results in only 12 volts of potential. A short circuit across E and F is the worst-case voltage difference of adjacent batteries in the upper circuit (24 volts). On the other hand, the connections are somewhat more complicated in the upper circuit, which means wiring mistakes are somewhat more likely. Neither circuit is "better" than the other; instead, these are shown here simply to advise the reader of potential issues to keep in mind when laying out the arrangement of a set of batteries.

The result of a short circuit turned this battery bank into an arc welder. Courtesy Tony Helmholdt.

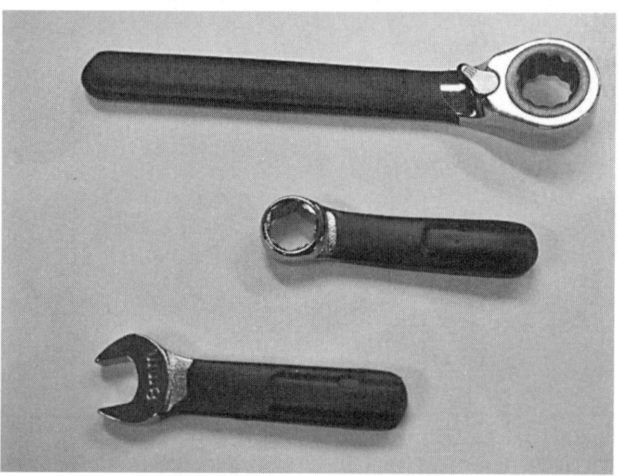

The handles of these wrenches have been coated with rubberized "dip" coating. Available at most auto parts stores, this type of electrically insulated coating can be applied to ratchets, piers, wrenches, and just about any other tool. Other EV builders have used large diameter, thick-walled heat shrink tubing to achieve similar protection on their tools. The key is to protect the entire length of the tool handle, ensuring there is no path for electricity to flow through and out the end of the tool. Courtesy Tony Helmholdt.

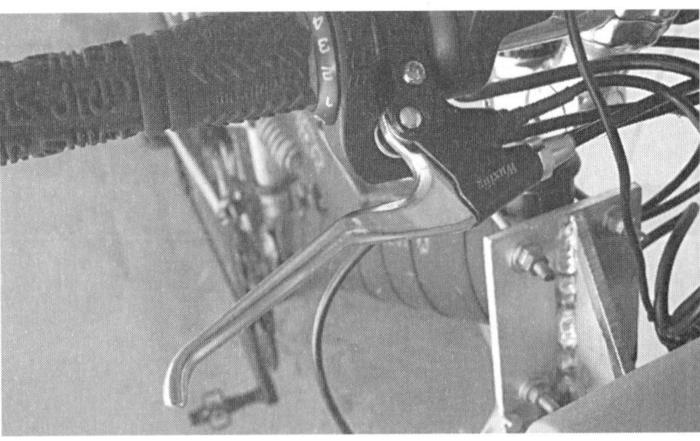

Safety is important even on low-power EVs, such as this electric bicycle that operates at 36 volts. Note the Wuxing interlocked brake lever. A microswitch inside brake lever housing is designed to disable the controller and kill power whenever the brakes are applied. Both front and rear brakes incorporate this safety feature, which is cheap insurance against having a runaway electric bicycle capable of 20+ mph.

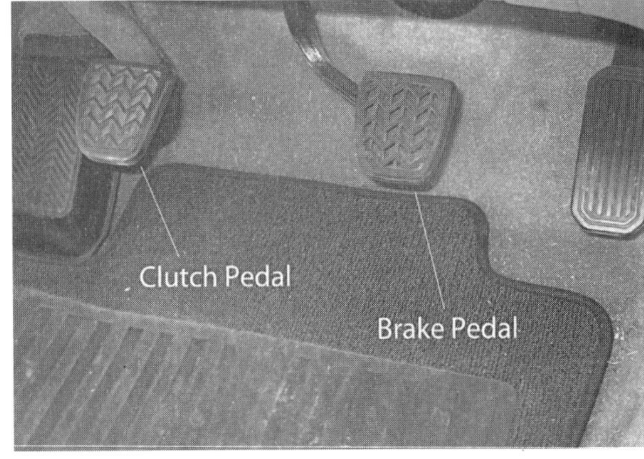

Some owners who have eliminated the clutch in their EVs still retain the clutch pedal as a safety device. By installing a relay switch activated by the movement of the pedal, a master power off contactor can be operated. In a panic situation, depressing the clutch kills power to the drives. Simple and effective.

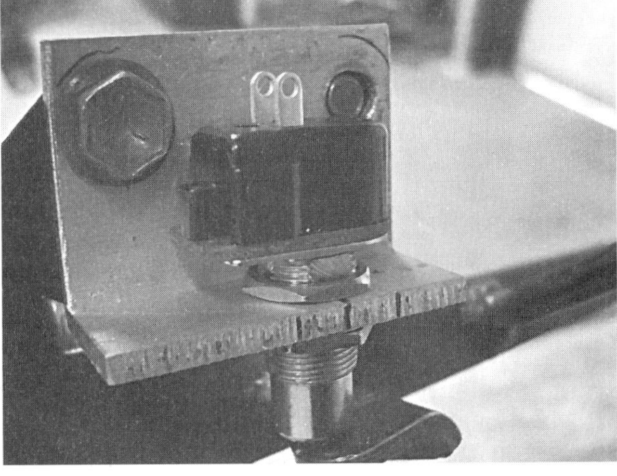

It is very easy to forget that batteries are always On. It's also easy to forget to turn off an EV, as they make little to no noise when activated. The owner of an electric motorcycle found out the hard way that his power system was activated when he inadvertently twisted the throttle one day when showing the bike to a group of people. The good news is the motorcycle didn't hit anyone when it took off at full speed. The even better news is the owner decided to take safety much more seriously from that day onward. Shown here is a momentary Off switch that activates whenever the kickstand is put down. When the motorcycle is parked it is impossible for power to flow from the battery to the controller. Simple, but very effective. Courtesy Tony Helmholdt.

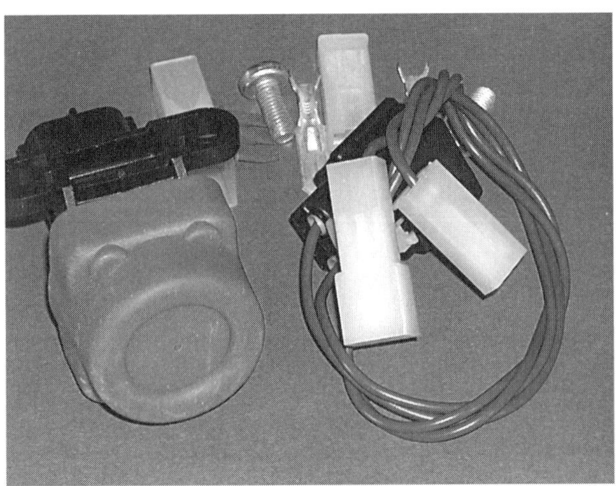

In the event of a vehicle crash, it is desirable to disconnect the main battery pack from the motor control system. Shown here is an "inertia" switch that can be used to switch off power to a main contactor. The unit is normally closed but opens the circuit upon a hard impact. Courtesy KTA Services.

locations, you should install individual circuit over-current devices (e.g., a fuse) to protect each bank.

Safe Construction, Charging, Testing and Shakedown

There are a myriad of ways you can hurt yourself, someone else, or the vehicle itself during construction of an EV. It's impractical to cover all possible scenarios here, and to that end the first word of caution is that you must use common sense when building, charging, and testing your EV. When in doubt about anything, stop and think. Don't perform any task until you've fully understood the steps required and the ramifications of what could happen if things go wrong. Keep reminding yourself that taking an extra fifteen minutes to do a job right now may save hours or days repairing the damage later if you make a mistake.

Another thing to consider is finding a partner to help. Two sets of eyes are always better than one on a problem. Talk things through with your partner. Discuss the plans for each garage session prior to starting work. Remind each other where the fire extinguisher is located and how to use it. Ensure that each of you is wearing long sleeve shirts, long pants, and closed-toe shoes. An ounce of prevention here can go a long way toward ensuring a safe session in the workshop.

Your work space should also be a safe environment in which to work. Things like personal safety equipment (e.g., safety glasses, gloves, hearing protection, etc.) and work space safety gear (e.g., electrical-rated fire extinguisher, an acid neutralizer, etc.) are absolutely required. Don't skimp on substandard tools, either; having the right tool for each job will not only make the tasks go easier but will result in fewer accidents and injuries, too.

When working on and around your vehicle, always assume every component is energized. Modified tools that are insulated and/or cut down to avoid spanning terminal distances are a good idea. Wear UL listed rubber gloves designed for electrical work. And always disconnect the battery bank whenever working on and around this engine bay. This can be done by many methods. For example, on some vehicles it's a simple matter to just disconnect one or both main leads to the battery bank. On others, it's better to remove the main fuse or flip the circuit breaker. Another good method is to have a master OFF switch that breaks the main power feed line via a switched fail-safe contactor.

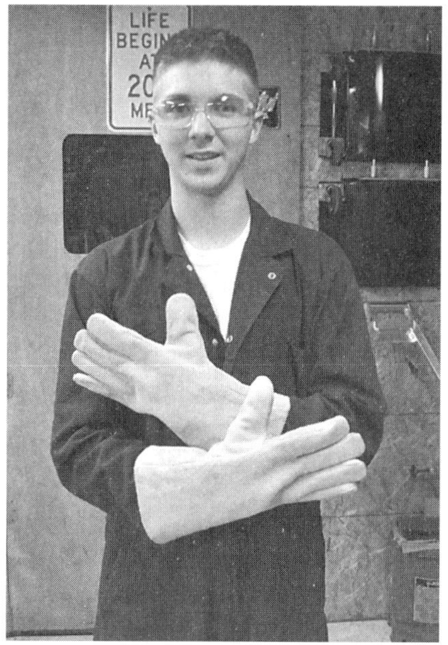

Get into the habit of wearing protective clothing when working on and around your EV, even if it's just to check the water level in your batteries. Long rubber insulated approved lineman's gloves, long sleeved shirts, rubber-soled boots, and safety glasses are more important than you may realize. Courtesy Tony Helmholdt.

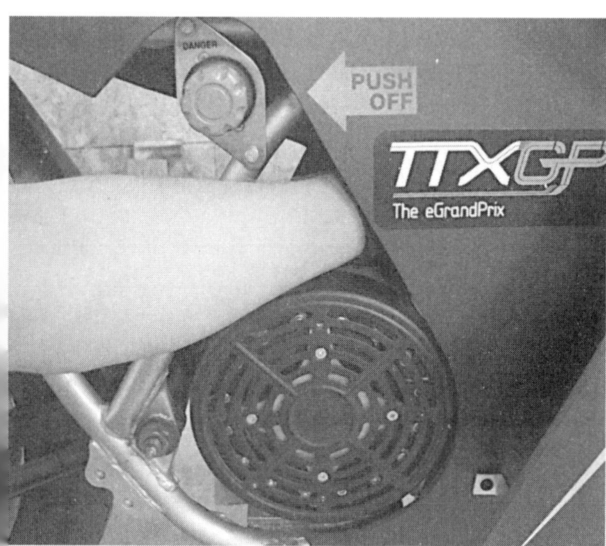

The good news is the large master Off switch that is easily accessible—and therefore will likely get used by the owner when servicing the vehicle. EVs make very little noise when switched on, and it's all too easy to forget that the circuits are armed and ready. Now, the bad news is that the owner is reaching into a "mystery" hole, where it is easy to get electrocuted. Be extremely careful when working in areas like this, and be aware of what you are touching and what is around you and the tools you are using. Courtesy Tony Helmholdt.

It's important to have a reliable means of putting out a fire in your garage if one should start. Fire extinguishers are divided into categories based on the types of fires they can stop. For example, Class A extinguishers are for ordinary combustible materials such as paper, wood, cardboard, and most plastics. Class B fires involve flammable or combustible liquids such as gasoline, kerosene, grease and oil. Class C fires involve electrical equipment, such as appliances, wiring, circuit breakers and outlets. Some extinguishers, such as those shown here, are multi-rated, meaning that they can be used effectively on type A, B, and/or C fires.

Volts or Amps: Which Is More Dangerous?

Electrical vehicle enthusiasts often debate which is the larger danger: voltage or current? Conventional wisdom states that it's the current coursing through a body that causes any actual internal damage. While this is true, it's not really the whole story.

The U.S. Department of Labor's Occupational Safety & Health Administration (OSHA) guidelines state there are three primary factors that affect the severity of an electrical shock a person receives when he or she is a part of an electrical circuit: (1) the amount of current flowing through the body (measured in amperes); (2) the path of the current through the body; and (3) the length of time the body is in the circuit. In addition, the severity of the shock is also affected by the voltage of the current, the amount of moisture present, and the general health of the person prior to the shock.

The effect of a shock can range from a barely perceptible sting or tingle to severe burns, cardiac arrest, and even death. The OSHA chart below shows the probable effect on a human body when subjected to different amounts of electrical current.

As you can see from the chart, electrical current flowing through the body is a serious danger, but it's important to remember that Ohm's law says that volts and amps are related by resistance; in this case, the resistance of the human body. In dry conditions, human skin is relatively non-conductive, with a resistance value of approximately 100,000 ohms. This means that the electrical current that can pass through the skin and into a body when subjected to 120 volts is roughly: I = V ÷ R = 120V ÷ 100,000 ohms = 0.001 amp. This small amount would likely just result in a slight tingling sensation or "sting."

In wet conditions, however, things change dramatically. The resistance of wet skin can be as low as 1,000 ohms or less. Applying the same 120 volts to a wet body would result in 120V ÷ 1,000 ohms = 0.12 amperes flowing into the body. The OSHA chart below shows that this can result in extreme pain, respiratory arrest, and severe muscular contractions. Death is even possible at this level.

It's also important to note that the internal resistance of a human body is even lower than the wet skin resistance. The actual resistance depends upon a number of factors, but an average value that scientists use is somewhere in the range of 500 ohms or less. In other words, if human skin is punctured and electricity applied directly to the wound, the result can be overwhelmingly significant. Currents of 2 amps (12V ÷ 500 ohms) or more is possible. Said another way, electricity applied to an open wound can cause the heart to stop and the person to die.

Scary, isn't it? Well, yes; electricity is dangerous if mistreated. And this is why it's so important to exercise extreme caution whenever working around electrical circuits. Even low-voltage circuits. Electricity can injure and kill human beings quite easily if things go amiss. It really doesn't matter if it's the voltage or the current in the circuit; they're both related and can both create dangerous electrical shocks.

Current [amps]	Probable Effect On Human Body
0.001	Human perception level. Slight tingling or stinging sensation. Dangerous under certain conditions.
0.005	Slight shock felt; not painful but disturbing. Average individual can let go of conductor. However, strong involuntary reactions to shocks in this range may lead to injuries.
0.006 to 0.03	Painful shock, muscular control is often lost. Person may or may not be able to let go of conductor.
0.05 to 0.15	Extreme pain, respiratory arrest, severe muscular contractions. Individual cannot let go. Death is possible.
1 to 4.3	Ventricular fibrillation. Muscular contraction and nerve damage occur. Death is likely.
10	Cardiac arrest, severe burns and probable death

GLOSSARY

A

A: *See* Ampere

Absorbed Electrolyte: A type of electrolyte used in absorbed glass mat batteries.

Absorbed Glass Mat (AGM): A type of sealed lead-acid battery in which the electrolyte is absorbed and retained in a matrix of glass fibers. These types of batteries tend to have low internal resistance, good power characteristics, and good charging characteristics and tend not to spill in accidents. Most AGM-type batteries can be installed in any orientation, even upside down.

AC: *See* Alternating Current

A/C: A common abbreviation for vehicle air-conditioning.

Activation: The act of adding electrolyte to a dry battery, thereby making the battery functional.

Active Metals: Metallic elements that react with each other. For example, in a lead-acid battery the two active metals are lead peroxide (positive) and spongy lead (negative).

Adjustable Speed Drive: A device that controls the speed and efficiency of DC electric motors, usually by way of rectifying AC power to DC, and then controlling the voltage output. Also known as DC Drive.

AGM: *See* Absorbed Glass Mat

Ah: *See* Ampere-hour

Alkaline Battery: A type of non-rechargeable battery used in consumer appliances.

Alternating Current (AC): An electric current that periodically reverses direction, flowing forward, then backward, over and over again in rapid succession. In the U.S., most household current is alternating current that reverses direction sinusoidally at 60 cycles per second, or Hertz. In Europe, household current is also alternating, but operates at 50 Hz.

Ambient Temperature: The average temperature of the surrounding medium, typically air, in which an electrical device is operated. Often used to help specify a battery's performance at a given temperature.

Ammeter: A device used to measure the current in an electrical circuit. Most ammeters express the current in units of amperes. In low-current ammeters, the current is often expressed in milli-amperes (1/1000 of an amp).

Amp: *See* Ampere

Ampere: A unit of electrical current. Amperes are related to voltage and resistance by Ohm's law.

Amplitude (I, A, Amp): The maximum height or value above the zero point of a wave. Often used to express the maximum voltage peak in a sinusoidal AC waveform.

Amp-hour: *See* Ampere-hour

Ampere-hour (Ah): A unit of electrical energy. One amp-hour is equivalent to one ampere of electrical current flowing for one hour. Often used as a measure of the energy provided to or drawn from a battery.

Anode: The negative electrode or terminal of a battery.

Automotive Battery: *See* Starting, Lighting, and Ignition Battery

Automotive Post: A type of battery terminal commonly found on starting batteries used on internal combustion engines. A cable lug fits around the post to make the connection.

B

Back EMF: A negative voltage or potential that drives backward against the normal flow of electricity in a circuit. Electrical motors often create back EMF at high speed that slows the normal flow of current to the device; back EMF increases with motor speed, thereby reducing the motor's ability to turn faster.

Battery: A collection of electro-chemical galvanic cells, typically connected in series and packaged into a single body, that is used to store chemical energy. The battery provides a voltage difference between its two terminals, allowing a flow of electrons through an external closed circuit, thereby converting chemical energy into electricity.

Battery Council International (BCI): A trade association for the lead-acid battery industry that researches batteries, develops and publishes technical data, establishes engineering standards, and promotes the interests of the lead-acid battery industry.

Battery Cycle: A process in which a battery is fully charged and then discharged (or vice versa).

Battery Cycle Life: The number of charge-discharge cycles that a battery can undergo before degrading below its capacity rating.

Battery Electric Vehicle (BEV): A type of electric vehicle that uses chemical energy stored in rechargeable battery packs, along with an electric motor and controller, to provide propulsion to the vehicle.

Battery Management System (BMS): An electronic system used to monitor (and often protect) a vehicle's battery system. Often incorporated into the battery charging system, especially with lithium-based batteries.

Battery State of Charge: *See* State of Charge

Battery Terminal: The electrical connection point or contact on a battery used for connecting the battery to a cable or wire. Battery terminals come in a variety of styles and designs, including Automotive-post, L-post, and Universal-post.

BEV: *See* Battery Electric Vehicle

BCI: *See* Battery Council International

BLDC: *See* Brushless DC Motor

BMS: *See* Battery Management System

Brushed DC Motor: A type of electric motor that uses brushes (often made of carbon) to conduct electricity from the non-rotating stator to the moving rotor, thereby allowing the rotor windings to energize and create a magnetic field.

Brushless DC Motor (BLDC): A type of electric motor that employs permanent magnets in the rotor to react against a magnetic field created in the windings of the non-rotating stator. An independent means of

commutating (i.e., rapidly switching) the flow of electricity through the windings is required, which is typically performed by way of transistors, an optical encoder, hall-effect encoder, and/or similar electronic device.
BSOC: An abbreviation for Battery State of Charge. *See* State of Charge

C

C#: A description of a battery's rate of discharge (e.g., C20). The number following the letter C indicates the number of hours to fully discharge the battery at a constant current.
C-Rate: A measure of battery current expressed in terms of the battery's capacity. For example, a 10 amp-hour battery with a current of 10 amps would be rated at 1C. At 20 amps, the battery would be 2C, and so on. C-rate is used to rate both charging and discharging.
CA: *See* Cranking Amps
Capacitor: A device that can store electrical charge. Capacitance is typically measured in units of Farads or, more commonly, fractions of a Farad (e.g., micro-Farad, or one millionth of a Farad).
Capacity: The energy content of a battery, typically expressed in units of ampere-hours. Capacity is the total number of ampere-hours that can be withdrawn from a fully charged battery under specific conditions of discharge. Typically, the capacity of a battery is determined by measuring the time it takes to discharge at a constant current until a specified lower-bound voltage is reached.
CARB: An abbreviation for the California Air Research Board, which is a department within the California Environmental Protection Agency, whose goals include attaining and maintaining healthy air quality, protecting the public from exposure to toxic air contaminants, and providing innovative approaches for complying with air pollution regulations and rules.
Cathode: The positive electrode or terminal of a battery.
CCA: *See* Cold Cranking Amps
CE: *See* Commutator End
Cell: *See* Electrochemical Cell
Cell Mismatch: A condition where cells within a battery (or batteries within a bank of batteries) contain different voltage levels and/or capacities.
Charge: The process of adding energy to a battery, thereby replenishing its electrical charge.
Circuit: A pathway through which an electrical current can flow. The term circuit is typically used to represent a collection of wires, resistors, batteries, motors, and so forth, although any pathway through which electricity flows is technically a circuit.
Circuit Breaker: A safety device used to protect against over-current flow situations. Circuit breakers are most commonly a mechanical device that automatically opens (i.e., breaks, or shuts off) the circuit if the current exceeds a certain ampere value for a certain duration.
Cold Cranking Amps (CCA): A measure of performance of starting-type batteries. The current that a new and fully charged battery can deliver continuously for thirty seconds at an operational temperature of 0 degrees F, while maintaining a voltage of 1.20V per cell.
Commutator End (CE): The end of an electric motor that contains the brushes and commutator. Some motors have drive shafts that extend from this end of the motor.
Conductor: A wire, cable, bus bar, or other conductive item through which electrical current can flow.
Contactor: An electo-mechanical switch that is opened (and closed) by way of turning an electromagnetic coil on (or off). Contactors are often used in EV applications to switch on high-voltage/high-current circuits via a low-voltage/low-current circuit and remote switch.
Controller: A motor speed controller.
Coulomb: A unit of electrical charge equal to the amount of electricity transferred by a current of one ampere for one full second.
Cranking Amps (CA): A measure of the performance of starting-type batteries. The current that a new and fully charged battery can deliver continuously for thirty seconds at an operational temperature of 32 degrees F, while maintaining a voltage of 1.20V per cell.
Current: The rate of flow of electrical charge. Typically measured in units of amperes.
Current-Limiting Charger: A battery charger that maintains the charging current at a constant value during the charge process (but allows the voltage to fluctuate as required). Often used with Nickel-Cadmium and Nickel Metal-Hydride batteries.
Cycle: The complete process of a battery being fully charged and then discharged.
Cycle Life: The number of charge-discharge cycles a battery can undergo before losing its ability to hold a useful charge. Cycle life is typically dependent upon the depth of discharge (DOD) of the battery.

D

DC: *See* Direct Current
DC-DC Converter: An electrical device used to step-down an electric vehicle's primary operating voltage (e.g., 144 volts) to a lower voltage (e.g., 12 volts) to operate the vehicle's accessories, such as lights, stereo system, horn, etc.
Deep-Cycle Battery: A battery designed to deliver a near-constant voltage as the battery discharges. Deep-cycle batteries can be routinely discharged to 80% or more of their capacity before being recharged.
Depth of Discharge (DOD): A measure of the amount of energy that has been removed from a battery, typically expressed as a percentage of the total capacity of the battery. For example, a 60% depth of discharge of a battery means that 60% of the energy in the battery has been used, and the battery now only holds 40% of its full charge.

Direct Current (DC): An electrical current that flows in one direction. Direct current is produced by sources as batteries and photovoltaic solar panels.
Discharge: The conversion of chemical energy stored in a battery to that of electrical energy.
DOD: *See* Depth of Discharge

E

E-bike: An electric-powered motorcycle, scooter, or bicycle.
Efficiency: The ratio of the output of a device to the input. For example, if an electric motor requires 100 watts input but only produces the equivalent of 80 watts of power at its output shaft (i.e., 20 watts were consumed as heat, friction, and other "losses"), the motor is said to be 80/100 = 80% efficient.
Electric Vehicle (EV): A vehicle powered by way of an electric motor. While normally used to refer to electric-powered automobiles, the term *electric vehicle* also encompasses trucks, motorcycles, scooters, golf carts, forklifts, trolleys, burden carriers, and other devices.
Electrochemical Cell: A device used for generating an electromotive force (i.e, voltage) and electrical current from chemical reactions. Electrochemical cells are the building blocks of a battery.
Electrode: A conductor through which electrical current enters or leaves a battery (or similar device).
Electromotive Force (EMF): The electric potential difference created by separating positive and negative charges, thereby generating an electric field. *See also* Voltage
Electrolyte: A chemical compound (typically a liquid) that ionizes, thereby providing an electrically conductive medium. An electrolyte is a non-metallic conductor of electricity in which the flow of electricity is by way of ions. In a typical flooded lead-acid batter, the electrolyte is a solution of sulfuric acid.
Electron: A negatively charged particle that orbits the nucleus of an atom.
ElectroSource: A manufacturer of lead-acid batteries.
EMF: *See* Electromotive Force
Energy: The capacity of a physical system to do useful work. Energy can take many different forms, such as kinetic, potential, spring, or chemical. For discussions of electrical circuits, energy is often expressed in terms of watt-hours or similar units.
Energy Density: A measure of the amount of energy contained in a specific amount of a substance. Often stated in watt-hours/lb or watt-hours/kilogram.
Equalizing Charge: A voltage applied to the batteries in a battery bank (or the cells of a battery) to balance them out and bring them all back to the same nominal voltage level. Often used in lead-acid battery banks.
EV: *See* Electric Vehicle.
EV Grin: The look on a person's face after their first ride in an electric vehicle.
Exercise: One or more discharge and recharge cycles applied to a battery.

F

Fast Charge: A technique in which a battery is recharged at a rate faster than is typical.
FCEV: *See* Fuel Cell Electric Vehicle
FCV: *See* Fuel Cell Electric Vehicle
Float Charge: A charging current applied to a battery that is intended to overcome the battery's self-discharge rate. A charging current used to maintain a battery in a fully charged state. *See also* Trickle Charge
Flooded Battery: *See* Flooded Cell
Flooded Cell: A type of battery in which the electrolyte is a liquid solution, typically acid. For example, a flooded lead-acid battery will have a liquid solution of sulfuric acid used as the electrolyte.
Flywheel: A device that uses the angular momentum of a spinning wheel or mass to store kinetic energy, which then can be extracted as electrical energy by way of a generator attached to the output shaft of the flywheel.
Frequency: The number of complete cycles per second of an alternating current. Usually expressed in units of Hertz, or Hz. The standard frequency of normal household electric power in the U.S., for example, is 60 Hz, or 60 cycles per second.
FSM: Abbreviation for Factory Service Manual.
Fuel Cell: An electrochemical energy conversion (and delivery) device in which electricity is produced directly from a fuel (e.g., hydrogen) and an oxidizer (e.g., oxygen). Fuel cells are different from electrochemical batteries in that they consume a fuel (and oxidizer) from an external source and therefore need to be periodically replenished. In contrast, batteries are technically just energy storage devices.
Fuel Cell Electric Vehicle (FCEV) (FCV): An electric vehicle that is provided electricity by way of a fuel cell.
Full Hybrid: A hybrid electric vehicle whose electric motor is powerful enough to move the vehicle without assistance from the internal combustion engine, and vice versa.
Fuse: An over-current protection device. Most fuses operate by way of a conductor that melts when the current pass through it exceeds a certain ampere value for a specific duration of time. Upon melting, the conductor opens the circuit, thereby preventing any more flow of electricity.

G

Gassing: The process of gas creation during charging of a battery. For example, when charging lead-acid batteries, hydrogen gas is produced (which requires safety precautions to ensure proper ventilation and avoid the danger of explosion.)
Gel Cell: A type of battery in which the electrolyte is in a gel form; i.e., non-liquid.
Genesis: A popular brand of lead-acid battery often used in electric vehicles.
GNB: A manufacturer of lead-acid batteries.
Grid: A term used to describe a network of electrical generation,

transmission, and distribution lines. In electric vehicle discussions, *the grid* usually refers to electrical power delivered from a power or utility company to a house or building.
Ground: An electrically conductive connection between an electrical circuit or device and earth.
Group Size: A standardized measure of a battery's external dimensions. For example, all Group 25 batteries, regardless of manufacturer, will have external dimensions of 230mm x 175mm x 225mm; weights and capacities may vary between manufacturers for a given Group Size, but external dimensions will be identical.
Gross Vehicle Weight Rating (GVWR): The maximum allowable design load for a particular vehicle. It includes the vehicle weight and the max payload, including the weight of passengers. When selecting a vehicle for conversion, the GVWR is an important number to keep in mind, because the weight of batteries (especially lead-acid types) is substantial.

H

H_2SO_4: Sulfuric Acid
Hawker: A manufacturer of lead-acid batteries.
Heat Sink: A device used to draw heat energy out of an electrical component. For example, motor controllers are often attached to large pieces of thermally conductive aluminum fitted with cooling fins. As the controller warms up, heat is transferred to the aluminum plate, where it is dissipated via convection to the surrounding air.
Hertz: A unit of measurement used to express frequency. One Hertz is equal to one cycle per second.
HEV: *See* Hybrid Electric Vehicle
hp: *See* Horsepower
Horizon: A brand of lead-acid battery often used in electric vehicles.
Horsepower (hp): A unit of power, which is a measure of the rate at which work is performed. One horsepower is equal to 746 watts, or 550 lb-ft/second.
Hybrid Electric Vehicle (HEV): A vehicle propelled by two (or more) sources of power, one of which is electrical power. Most commercial hybrid electric vehicles combine an internal combustion engine with an electric motor. When at rest, both the electric motor and internal combustion engine are switched off. During acceleration and for short periods of constant velocity, the electric motor is used to provide propulsion. At higher speeds and for trips that are relatively long distance, the internal combustion engine takes over from the electric motor to power the vehicle.
Hydrogen: A colorless, odorless gas that can be given off at the negative plate of a lead-acid battery during charging. Hydrogen gas is considered explosive and must be treated accordingly.
Hydrogen Fuel Cell: A fuel cell that utilizes hydrogen (and oxygen) to produce electricity (and other byproducts, such as water). *See also* Fuel Cell.
Hydrometer: A specialized tool for measuring the specific gravity of a liquid. Hydrometers are often used to test the electrolyte in a flooded cell-type battery.
Hz: *See* Hertz

I

ICE: *See* Internal Combustion Engine
IEEE: An acronym for the Institute of Electrical and Electronics Engineers.
Immobilized Electrolyte: A battery electrolyte that is immobilized, or held in place, against the internal battery plates; i.e., not a free-flowing liquid. For example, an immobilized electrolyte is often used in gel cell batteries.
Impedance: A measure of resistance to the flow of electrical current that is created by both Ohmic resistance and reactance.
Inductor: An electrical device that stores or transports electrical current as a magnetic field.
Insulator: A non-conducting material used to isolate electrical charge.
Internal Combustion Engine (ICE): A device that uses combustion of a fuel (and oxidizer) to perform mechanical work. In electrical vehicle discussions, an internal combustion engine typically refers to the vehicles original gasoline- or diesel-powered engine. Most vehicles driven on the road today are ICE-powered vehicles.
Internal Resistance: The resistance to the flow of electrical current inside a battery. Typically measured in units of ohms.
Inverter: An electrical device that converts DC power (or current) to AC power (or current).
Ion: A particle that is electrically charged (positive or negative); an atom or molecule or group that has lost or gained one or more electrons.

J

Joule: A unit of work or energy equal to one watt dissipated for one second. One kilowatt hour is equivalent to 3,600,000 Joules.
Jumper: A short length of conductor or wire used to temporarily connect two parts of an electrical circuit.

K

Kilo: A metric prefix for 1000. For example, one kilovolt is equal to 1000 volts.
Kilowatt (KW): A measure of electrical power equal to 1000 Watts.
Kilowatt-Hour (KWhr): A unit of energy or work.
KW: *See* Kilowatt
KWhr: *See* Kilowatt-Hour

L

L-Post: A type of battery terminal that is, literally, shaped like the letter L. A bolt or screw through a hole in the flat part of the L connects the terminal to a cable lug.
Lead (Pb): A chemical element with the atomic number of 82. Used in lead-acid batteries.
Lead-Acid Battery: A type of battery in which electrodes of lead oxide and metallic lead are separated by a sulfuric

acid electrolyte. Lead-acid batteries are a very mature technology that offer rugged and very reliable performance. Due to their relatively low cost and high energy density, lead-acid batteries are the most common form of batteries used in electric vehicle conversions.

$LiFePO_4$: *See* Lithium Iron Phosphate battery

Li-Ion: *See* Lithium Ion

Lithium: A chemical element with the atomic number of 3. Lithium is a soft alkali metal a silver-white color that under standard conditions is the least dense solid element and the lightest weight metal.

Lithium-Ion (Li-Ion): A type of battery in which a lithium ion moves between the anode and cathode. Lithium-Ion batteries are a relatively new but rapidly expanding technology that offer high-energy density and low weight. Used in many portable consumer electronic products and are becoming more and more prevalent in electric vehicle conversions.

Lithium Iron Phosphate ($LiFePO_4$): A type of rechargeable battery used in high-end EV conversion projects. Compared to standard lithium-ion batteries, these lithium iron phosphate batteries do not have any volatile thermal issues, therefore making them safer to use in high-power applications.

Lithium Polymer: A type of battery in which a lithium electrolyte is contained in a solid polymer composite.

Load: A piece of equipment that uses electricity for power. For example, a motor is considered to be a load.

Load Current: The discharge current from a battery when subjected to a closed circuit load.

M

Maintenance Requirement: The maintenance needs of a battery to promote long battery life. For example, flooded lead acid batteries require topping off of fluids, periodic equalizing charges, etc. Similarly, nickel-based batteries often require periodic full discharge-recharge cycles to eliminate memory effects. Other batteries, such as lithium-ion, have relatively low maintenance requirements during the course of their useful lifetimes.

Memory Effect: A term that refers to the loss of capacity in a battery due to non-optimal discharging. For example, nickel-based batteries often require periodic full discharge-recharge cycles to minimize memory issues.

Mho: A unit of electrical conductivity equal to the reciprocal of resistance. Mho is the reverse spelling of Ohm.

Milli: A metric prefix meaning 1/1000. For example One millivolt is equal to 0.001 volts.

Milliampere: One one-thousandth of an Ampere.

NEV: *See* Neighborhood Electric Vehicle

Negative Terminal: The battery terminal from which current flows to an external load when the battery is discharging.

Neighborhood Electric Vehicle (NEV): A classification for small electric vehicles designed for low-speed operation in restricted areas, such as a retirement community. Neighborhood electric vehicles are often exempt from some Department of Transportation and/or Motor Vehicle Department safety standards.

NiCad: A registered trademark of SAFT Corporation for a brand of Nickel-Cadmium batteries. Commonly (and erroneously) used to describe all Nickel-Cadmium batteries.

NiCd: *See* Nickel-Cadmium

Nickel-Cadmium (NiCd): A type of battery that uses separated nickel oxide hydroxide and metallic cadmium as electrodes and uses an alkaline electrolyte, such as potassium hydroxide. Nickel-cadmium batteries are typically used where long life, high discharge rates, and extended temperature ranges are important. Nickel-cadmium batteries are a popular and mature technology but are slowly being overtaken in the marketplace by nickel-metal hydride and lithium-ion batteries, due in large part to issues of energy density, cost, and environmental concerns of the toxic cadmium used in the battery.

Nickel-Metal Hydride (NiMH): A type of battery that uses separated nickel oxide hydroxide and a hydrogen absorbing alloy for the electrodes and uses an alkaline electrolyte, such as potassium hydroxide. Nickel-metal hydride batteries have relatively high energy density capabilities but come at the expense of reduced cycle life. Nickel-metal hydride batteries are often viewed as a stepping-stone to lithium-based battery systems of the future. This is the type of cell used in the Insight's 144V battery pack.

NiMH: *See* Nickel-Metal Hydride

Noise: An electrical term meaning unwanted electrical signals, current, or voltage in an electrical circuit.

Nominal Voltage: A standard cell or battery voltage.

O

OEM: An abbreviation for Original Equipment Manufacturer. In EV discussions, OEM typically refers to the original factory constructor of the vehicle, e.g., Ford, Toyota, etc.

Ohm: A unit of electrical resistance. If one volt is applied to an electrical circuit that has one ohm of resistance, exactly one ampere of electricity will flow through the circuit.

Ohmic Resistance: A resistance to the flow of electrical current and one that is void of reactance. Typically measured in units of ohms.

Ohmmeter: A device used to measure electrical resistance.

Ohm's Law: The relationship between voltage, resistance, and current. Ohm's law states that Voltage = Current x Resistance.

Open Circuit Voltage: The difference in potential between the positive and negative terminals of a battery when the circuit is not connected to a load (or open).

Opportunity Charging: A term used to describe charging of an electrical vehicle at a location that is somewhere other than the vehicle's home base location. For example, an EV may be "opportunity charged" at the owner's place of occupation during the day. Similarly, an EV may be "opportunity

charged" at the grocery store while the owner is shopping. Opportunity charging is not typically meant to fully charge the vehicle, but rather to "top off" the batteries to increase the range.
Optima: A brand of sealed lead-acid battery often used in electric vehicles.
Overcharge: The act of charging a battery up to and beyond the point at which it is fully charged. Typically, overcharging a battery results in a condition where the battery can no longer absorb the charge and instead begins to heat up.
Ovonics: A manufacturer of nickel-metal hydride batteries.

P

Passivation Layer: A resistive layer that forms inside batteries on their internal cells after periods of prolonged storage, thereby affecting battery performance. Some batteries that have experienced passivation may require charge-discharge cycles to degrade the passivation layer and prepare the battery for usage.
Pb: *See* Lead.
PbA: An abbreviation for a lead-acid battery.
Permanent Magnet Motor: *See* Brushless DC Motor
Peukert Equation: A formula that quantifies the difference in battery capacity as a function of the rate at which the battery is discharged. Typically, as discharge rates increase, the effective capacity of a battery is reduced.
Peukert Number: An exponent in the Peukert equation that gives a measure of how well a battery performs at high discharge rates. Values close to 1 indicate that a battery performs relatively well, while higher numbers indicate reduced capacity and poorer performance at higher discharge rates. Peukert numbers are determined via empirical test of a battery.
PEV: An abbreviation for Personal Electric Vehicle.
Polymer: In electrical vehicle discussions, a polymer usually refers to an electrical insulator used inside of a battery that allows the passage of ions through its medium.
PHEV: *See* Plug-in Hybrid Electric Vehicle
Photovoltaic Cell (PV): A solar cell that converts light directly into electricity. When combined with other photovoltaic cells, a solar panel, module, or array can be constructed to provide electricity.
Plug-in Hybrid Electric Vehicle (PHEV): A hybrid electric vehicle, whose battery can be recharged by an external source (e.g., via household electricity). Plug-in electric hybrid vehicles often have the ability to move in pure electric mode for finite distances, without any application of the internal combustion engine. *See also* Hybrid Electric Vehicle
PMDC: An abbreviation for Permanent Magnet DC motor.
Polarity: The condition of being either positively or negatively charged.
Potentiometer: A variable resistor often used as the throttle, which varies the speed output of a motor controller. Also informally known as a *pot.*
Potential: *See* Voltage
Power: The rate at which work is performed, or the rate at which energy is transferred. Electrical power is commonly measured in units of Watts. One horsepower is equivalent to 746 watts.
Power-Sonic: A manufacturer of lead-acid batteries.
Primary Battery: A battery made up of primary cells (i.e., non-rechargeable cells). A non-rechargeable battery. A battery that undergoes an irreversible change during discharge. A standard "AAA" battery that one purchases at a hardware or grocery store is typically a primary battery and must be thrown out or recycled after use.
Primary Cell: An electrochemical cell that cannot be recharged via the application of electricity.
Protection Circuit: Electronic circuitry built into a battery pack to maintain safety (e.g., from overcharging or overheating).
Pulse Width Modulation (PWM): A modern technique that uses electronic circuitry for controlling voltage to (and hence, speed of) an electric motor. Pulse width modulation involves the rapid cycling on and off of a power supply to essentially regulate the delivered voltage to a device, such as a motor.
PV: *See* Photovoltaic Cell
PWM: *See* Pulse Width Modulation

Q

Quick Charge: *See* Fast Charge

R

Range: The distance that an electric vehicle can travel without its batteries being recharged.
Rapid Charge: *See* Fast Charge
Rated Capacity: The capacity of a battery. Typically Rated Capacity refers to the number of ampere-hours a battery can deliver under a specific set of conditions.
Reactance: The resistance to the flow of electrical current by way of inductive and/or capacitive resistance.
Rechargeable Battery: A type of secondary cell or battery in which the charge can be restored to full or partial charge by the application of electrical energy, such as through a battery charger. *See also* Secondary Battery
Rectifier: An electrical device that converts AC power to DC power.
Red Top Battery: An informal, or slang, term for a type of sealed lead-acid battery manufactured by Optima. Red top batteries are not classified as traction batteries and therefore are not typically used in traditional electric vehicle applications.
Regen: *See* Regenerative Braking
Regenerative Braking (Regen): An electro-mechanical system or process in which the kinetic energy of a decelerating vehicle is converted back into electrical energy and stored (typically via the battery) for future use, thereby (somewhat) increasing the vehicle's effective range and improving drivability. Regenerative braking is frequently found in AC-type electric vehicles but is much less common in DC-type electric vehicles.

Renewable Energy: Energy or power created from "natural" resources, such as solar, wind, or hydro-electric.
Reserve Capacity: A measure of performance for automotive starting-type batteries. Typically equal to the number of minutes a battery can be discharged at a specific current (e.g., 25 amperes) while maintaining a minimum terminal voltage (e.g., 1.75 volts per cell) at a specific ambient temperature (e.g., 80 degrees Fahrenheit).
Residual Capacity: The charge capacity remaining in a battery prior to it being recharged.
Resistance: The resistance or opposition to the flow of an electrical current. Typically measured in units of ohms.
Resistor: An electrical device that adds resistance to an electrical circuit.
RPM: An abbreviation for revolutions per minute.

S

Sag: The lowering of a voltage. In EV parlance, sag typically refers to drop in output voltage from a battery.
Sealed Battery: A battery type that does not allow venting of internal gases to the atmosphere. A battery that does not require the periodic addition of water or other liquids.
Secondary Battery: A battery made up of secondary cells (i.e., rechargeable cells). A rechargeable battery. Also known as a storage battery.
Sealed Lead-Acid Battery (SLA): A lead-acid battery that does not require the periodic addition of water.
Secondary Cell: A cell that can be recharged via the application of electricity.
Self-discharge: A battery that discharges due to an internal flaw.
SG: *See* Specific Gravity
Shield: A conductive sheath applied over the insulation of a wire or conductor for the purpose of reducing the induction of electrical noise into the circuit.
Short Circuit: An unwanted electrically conductive path or connection between two points in an electrical circuit.
Sine Wave: The normal waveform of a the typical AC voltage output.
SLA: *See* Sealed Lead-Acid Battery
SLI: *See* Starting, Lighting, and Ignition Battery
Spiral Wound: A type of lead-acid battery in which the "plates" are rolled up into a spiral (and often separated by a glass-mat-type separator. The Optima-brand batteries are an example of a Spiral Wound battery.
SOC: *See* State of Charge
Solar Cell: *See* Photovoltaic Cell
Specific Gravity (SG): The ratio of the density of a substance to that of water at a specific temperature and pressure (typically 39 deg F at sea-level pressure). The specific gravity of an electrolyte is often used to give an indication of the state of charge of the battery; i.e., the specific gravity of the electrolyte decreases with the battery's state of charge.
Stall: A condition when an electric motor is kept from rotating when voltage is applied. A motor's stall torque is normally the point of maximum torque output.
Standard Conditions: A widely recognized and specific set of temperature, humidity, pressure, and other variables under which a battery is tested.
Starter Battery: *See* Starting, Lighting, and Ignition Battery
Starting, Lighting, and Ignition Battery (SLI): A battery designed for use in a traditional internal combustion-powered vehicle. This type of battery is designed for rapid but shallow discharging, followed immediately by recharging by way of the engine's charging system (e.g., alternator).
Starved Electrolyte: A battery in which there is a minimal amount of electrolyte, thereby reducing the likelihood of out-gassing, and minimizing the amount checking and addition of water required. This term is often used when referring to absorbed glass mat, or AGM-type lead-acid batteries.
State of Charge (SOC): A measure of the amount of energy stored in a battery. Calculated by the ratio of the current charge to that of the maximum charged condition and expressed as a percentage.
Stationary Battery: A battery used in a fixed, non-mobile location, such as for use in a household photovoltaic power system.
Stink Motor: A slang term for an internal combustion engine.
Storage Battery: *See* Secondary Battery
Sulfation: A process in which lead sulfate crystals form on the internal plates of a lead-acid battery, thereby inhibiting current flow.
Sulfuric Acid: The primary acid compound used in lead-acid batteries. Chemical symbol: H_2SO_4.
Supercapacitor: A capacitor that can store (and release) a large amount of energy.
Surge: A momentary large increase in the current or voltage supplied in a circuit.
Switch: An electrical device used for completing, disconnecting, or changing connections in an electrical circuit, e.g. an on-off switch.

T

Terminal: *See* Battery Terminal
Thermistor: A temperature-sensitive resistor used to sense and measure temperature.
Thermostat: A temperature sensitive switch. Often used to control the operation of electro-mechanical device such as fans when the temperature rises or falls below a threshold value.
Topping Charge: A slow rate (i.e., low current) charge applied at the end of a fast-charge process applied to a battery. Often used on nickel-based batteries.
Traction Battery: A type of battery optimized for deep discharges. A battery used to power an electric vehicle.
Transformer: An electrical device consisting of multiple coupled windings that is used to transfer power by way of electromagnetic induction between circuits. Often used to step up or down voltage levels.

Trickle Charge: A means by which a battery is very slowly charged, often at the same rate at which the battery self-discharges, thus maintaining full capacity of the battery.
Trojan: A manufacturer of flooded cell lead-acid batteries. A popular brand of battery often used in electric vehicles, due in large part to its relatively low cost and high durability.

U

Ultracapacitor: *See* Supercapacitor
Universal Post: A type of battery terminal with a round post similar to the automotive post but with a threaded stud in the center of the post. A cable lug fits over the stud and a nut holds them together.
UPS: An abbreviation for Uninterruptable Power Supply, which is often a stationary battery used to provide emergency power during a power outage.

V

Valve Regulated Lead-Acid (VRLA): A type of lead-acid battery that is relatively low maintenance (e.g., they do not require the periodic addition of water). Also known as sealed lead-acid batteries (although they are not in fact normally fully sealed; i.e., there is usually a pressure relief/safety valve incorporated in the battery housing).
Variable Frequency Drive (VFD): A device used to control the speed (and efficiency) of AC motors.

VFD: *See* Variable Frequency Drive
Volt: The electrical "pressure" applied to a circuit by a battery, thereby driving a current. Also known as an electromotive force or potential.
Voltage (VA): The difference in electrical charge between two points in a circuit, typically expressed in units of volt-ampere.
Volt-Ampere(VA): A unit of measurement for the total amount of power that must be delivered to a load or device to enable it to function.
Voltmeter: An electrical device used for measuring voltage.
VRLA: *See* Valve Regulated Lead-Acid

W

Watering: The act of periodically adding water to a battery electrolyte to replace loss from evaporation and/or electrolysis.
Watt: A unit of power equal to 1 joule per second. The power dissipated by a current of 1 ampere flowing across a resistance of 1 ohm.
Watt-hour: A unit of energy equal to the power of one watt operating for one hour.
Wet Cell: A slang term for a flooded-cell type lead-acid battery

Y

Yellow Top: An informal or slang term for a type of sealed lead-acid battery manufactured by Optima. Yellow top batteries are designed for deep discharging and are often used in electric vehicle applications.

Z

Zapping: An informal or slang term used to describe the momentary application of a high current pulse to a battery being charged to improve its performance. Used primarily on nickel-based batteries.
Zero Emission Vehicle (ZEV): A vehicle that emits no tailpipe pollutants, such as an electric car.
Zero Crossing: The point on an AC wave where the wave crosses the zero-amplitutde line and changes sign from positive to negative (or vice versa).
ZEV: *See* Zero Emission Vehicle
Zinc-Air: A non-rechargeable-type of battery often used in small consumer applications, such as watches.

SOURCES

Vehicle Conversion Components & Kits

Alternative Vehicles Technology
www.avt.uk.com

Amped Bikes
ampedbikes.com

Canadian Electric Vehicles, Ltd.
canev.com

Electric Car Company of Utah
evequipmentsupply.com

Electric Vehicles of America, Inc.
evamerica.com

Electro Automotive
electroauto.com

EV Components, LLC
evcomponents.com

EV Parts
evparts.com

EV-Propulsion
ev-propulsion.com

EV Source
EVSource.com

Grassroots Electric Vehicles
grassrootsev.com

KTA Services, Inc.
kta-ev.com

Metric Mind Corporation
metricmind.com

Individual Components

Motors & Controllers

4QD
www.4qd.co.uk

AC Propulsion
acpropulsion.com

Advanced Motors & Drives
adcmotors.com

Alltrax
alltraxinc.com

Azure Dynamics
azuredynamics.com

Brusa
brusa.biz

AC Drive Systems & Components

Café Electric, LLC
cafeelectric.com
Zilla controllers

Curtis Instruments
curtisinstruments.com
Curtis controllers

D&D Motor Systems, Inc.
ddmotorsystems.com

Enertrac Corporation
enertrac.net
motorcycle hub motors

Kelly Controls, LLC
www.kellycontroller.com
Kelly motor controllers

Kostov Motors
kostov-motors.com
DC and AC motors

Lynch Motor Company
lemcoltd.com
permanent magnet DC motors

NetGain Motors
go-ev.com
WarP, ImPulse, and TransWarP motors

Piktronik
www.piktronik.com
AC & DC motor controllers

Sevcon
www.sevcon.com
AC & DC motors, controllers, chargers and converters

Batteries

Deka Batteries
East Penn Manufacturing
eastpenn-deka.com

Enersys Energy Products
enersys.com

Everspring Global, Ltd.
www.everspring.net
Thundersky lithium batteries

Exide Technologies
exideworld.com
Exide batteries

Optima Batteries
optimabatteries.com
Optima AGM batteries

Ping Batteries
pingbattery.com
Lithium iron phosphate batteries

Trojan Battery Company
trojanbattery.com
lead-acid batteries

U.S. Battery
usbattery.com

Battery Chargers, Accessories, & Instrumentation

AVCON
avconev.com

Current Logic
current-logic.com
DC-DC converters

Eagle Eye Power Solutions
eepowersolutins.com
digital hydrometer

EV Instruments
evinstruments.com

Battery Monitoring Systems

Iota Engineering
iotaengineering.com
battery chargers

Manzanita Micro Power Systems
manzanitamicro.com
battery chargers

Power Designers
powerdesignersusa.com
battery chargers, data loggers, conditioners

Russco
russcoev.com
battery chargers

Tecumseh Products/Masterflux
masterflux.com
Sierra air-conditioning compressors

ZAPI
zapi.co.za
battery chargers

Zivan
zivanusa.com
battery chargers, DC-DC converters

Other Useful & Interesting Websites

EV Album
EVAlbum.com
showcase for owner-built electric vehicles

Battery University
BatteryUniversity.com
a wealth of information on batteries and charging

Electric Auto Association
eaaev.org
dedicated to promoting electric vehicles

National Electric Drag Racing Association
nedra.com
dedicated to increasing awareness of EVs through racing

The Electric Vehicle Discussion List
evdl.org

Do It Yourself Electric Cars E-Mail List
hgroups.yahoo.com/group/diy_ev_cars

DIY Electric Car Forum
www.diyelectriccar.com/forums

Endless-Sphere Technology EV Forum
endless-sphere.com/forums

Pure Electric Vehicles
pureelectricvehicles.com
three-wheeled electric motorcycles

INDEX

ABOUT THE AUTHOR

Mark Warner is a professionally licensed mechanical engineer, a member of the Society of Automotive Engineers, and an expert in the fields of turbochargers and forced induction. He has over twenty-five years of machine design and fabrication experience, ranging from complex aerospace systems to high-performance automotive applications. He is the author of numerous patents and has published dozens of articles for the automotive press, including technical, how-to, and feature pieces. He is the author of *How to Keep Your Minivan Alive!*, *How to Keep Your SUV Alive!*, *Street Turbocharging*, and *Street Rotary*, also by HPBooks.

GENERAL MOTORS
Big-Block Chevy Engine Buildups: 978-1-55788-484-8/HP1484
Big-Block Chevy Performance: 978-1-55788-216-5/HP1216
Building the Chevy LS Engine: 978-1-55788-559-3/HP1559
Camaro Performance Handbook: 978-1-55788-057-4/HP1057
Camaro Restoration Handbook ('61–'81): 978-0-89586-375-1/HP758
Chevy LS Engine Buildups: 978-1-55788-567-8/HP1567
Chevy LS Engine Conversion Handbook: 978-1-55788-566-1/HP1566
Chevy LS1/LS6 Performance: 978-1-55788-407-7/HP1407
Classic Camaro Restoration, Repair & Upgrades: 978-1-55788-564-7/HP1564
The Classic Chevy Truck Handbook: 978-1-55788-534-0/HP1534
How to Rebuild Big-Block Chevy Engines: 978-0-89586-175-7/HP755
How to Rebuild Big-Block Chevy Engines, 1991–2000: 978-1-55788-550-0/HP1550
How to Rebuild Small-Block Chevy LT-1/LT-4 Engines: 978-1-55788-393-3/HP1393
How to Rebuild Your Small-Block Chevy: 978-1-55788-029-1/HP1029
Powerglide Transmission Handbook: 978-1-55788-355-1/HP1355
Small-Block Chevy Engine Buildups: 978-1-55788-400-8/HP1400
Turbo Hydra-Matic 350 Handbook: 978-0-89586-051-4/HP511

FORD
Classic Mustang Restoration, Repair & Upgrades: 978-1-55788-537-1/HP1537
Ford Engine Buildups: 978-1-55788-531-9/HP1531
Ford Windsor Small-Block Performance: 978-1-55788-558-6/HP1558
How to Build Small-Block Ford Racing Engines: 978-1-55788-536-2/HP1536
How to Rebuild Big-Block Ford Engines: 978-0-89586-070-5/HP708
How to Rebuild Ford V-8 Engines: 978-0-89586-036-1/HP36
How to Rebuild Small-Block Ford Engines: 978-0-912656-89-2/HP89
Mustang Restoration Handbook: 978-0-89586-402-4/HP029

MOPAR
Big-Block Mopar Performance: 978-1-55788-302-5/HP1302
How to Hot Rod Small-Block Mopar Engine, Revised: 978-1-55788-405-3/HP1405
How to Modify Your Jeep Chassis and Suspension For Off-Road: 978-1-55788-424-4/HP1424
How to Modify Your Mopar Magnum V8: 978-1-55788-473-2/HP1473
How to Rebuild and Modify Chrysler 426 Hemi Engines: 978-1-55788-525-8/HP1525
How to Rebuild Big-Block Mopar Engines: 978-1-55788-190-8/HP1190
How to Rebuild Small-Block Mopar Engines: 978-0-89586-128-5/HP83
How to Rebuild Your Mopar Magnum V8: 978-1-55788-431-5/HP1431
The Mopar Six-Pack Engine Handbook: 978-1-55788-528-9/HP1528
Torqueflite A-727 Transmission Handbook: 978-1-55788-399-5/HP1399

IMPORTS
Baja Bugs & Buggies: 978-0-89586-186-3/HP60
Honda/Acura Engine Performance: 978-1-55788-384-1/HP1384
How to Build Performance Nissan Sport Compacts, 1991–2006: 978-1-55788-541-8/HP1541
How to Hot Rod VW Engines: 978-0-91265-603-8/HP034
How to Rebuild Your VW Air-Cooled Engine: 978-0-89586-225-9/HP1225
Porsche 911 Performance: 978-1-55788-489-3/HP1489
Street Rotary: 978-1-55788-549-4/HP1549
Toyota MR2 Performance: 978-155788-553-1/HP1553
Xtreme Honda B-Series Engines: 978-1-55788-552-4/HP1552

HANDBOOKS
Auto Electrical Handbook: 978-0-89586-238-9/HP387
Auto Math Handbook: 978-1-55788-554-8/HP1554
Auto Upholstery & Interiors: 978-1-55788-265-3/HP1265
Custom Auto Wiring & Electrical: 978-1-55788-545-6/HP1545
Electric Vehicle Conversion Handbook: 978-1-55788-568-5/HP1568
Engine Builder's Handbook: 978-1-55788-245-5/HP1245
Fiberglass & Other Composite Materials: 978-1-55788-498-5/HP1498
High Performance Fasteners & Plumbing: 978-1-55788-523-4/HP1523
Metal Fabricator's Handbook: 978-0-89586-870-1/HP709
Paint & Body Handbook: 978-1-55788-082-6/HP1082
Plasma Cutting Handbook: 978-1-55788-569-2/HP1569
Practical Auto & Truck Restoration: 978-155788-547-0/HP1547
Pro Paint & Body: 978-1-55788-563-0/HP1563
Sheet Metal Handbook: 978-0-89586-757-5/HP575
Welder's Handbook, Revised: 978-1-55788-513-5

INDUCTION
Engine Airflow, 978-155788-537-1/HP1537
Holley 4150 & 4160 Carburetor Handbook: 978-0-89586-047-7/HP473
Holley Carbs, Manifolds & F.I.: 978-1-55788-052-9/HP1052
Rebuild & Powertune Carter/Edelbrock Carburetors: 978-155788-555-5/HP1555
Rochester Carburetors: 978-0-89586-301-0/HP014
Performance Fuel Injection Systems: 978-1-55788-557-9/HP1557
Turbochargers: 978-0-89586-135-1/HP49
Street Turbocharging: 978-1-55788-488-6/HP1488
Weber Carburetors: 978-0-89589-377-5/HP774

RACING & CHASSIS
Advanced Race Car Chassis Technology: 978-1-55788-562-3/HP562
Chassis Engineering: 978-1-55788-055-0/HP1055
4Wheel & Off-Road's Chassis & Suspension: 978-1-55788-406-0/HP1406
How to Make Your Car Handle: 978-1-91265-646-5/HP46
How to Build a Winning Drag Race Chassis & Suspension:
The Race Car Chassis: 978-1-55788-540-1/HP1540
The Racing Engine Builder's Handbook: 978-1-55788-492-3/HP1492

STREET RODS
Street Rodder magazine's Chassis & Suspension Handbook: 978-1-55788-346-9/HP1346
Street Rodder's Handbook, Revised: 978-1-55788-409-1/HP1409

ORDER YOUR COPY TODAY!
All books are available from online bookstores (www.amazon.com and www.barnesandnoble.com) and auto parts stores (www.summitracing.com or www.jegs.com). Or order direct from HPBooks at www.penguin.com/hpauto. Many titles are available in downloadable eBook formats.

R.C.L.

MARS 2012

A